Testudo kleinmanni
Photograph Franziska Biedenweg

You can love only that which you know,

and you will protect only that which you love.

Konrad Lorenz

Contents

Preface

I have been keeping various species of Mediterranean Tortoises in spacious outdoor enclosures with fixed shelters for more than forty years. The last few years furthermore enabled me to make experiences with the captive care of the African Pancake Tortoise, *Malacochersus tornieri*. Aside from these, I harboured the desire of also taking Egyptian Tortoises into my care, but remained hesitant for some time, because this species was generally believed to be rather demanding as to its captive requirements. I also found that it was fairly difficult to obtain detailed information on the ecology and biological needs of this species, which is, in my opinion, an absolute must before taking in any tortoise species. Following an exhaustive scouring of the existing literature and setting up everything I deemed necessary, five *Testudo kleinmanni* eventually moved in with me – and they captured my heart in no time.

In the process of exchanging experiences and observations with other keepers of this species, I made the accidental acquaintance of Franziska Biedenweg, who later confided in me that she was entertaining the idea of writing a book on the Egyptian Tortoise. This idea then turned into a joint project with the aim of making it easier for other interested people to obtain as much information as possible on *Testudo kleinmanni*. In the process, I found that it made sense to also draw from my many years of experience with the keeping of other species of *Testudo*.

The present book was created with the goal of contributing a little to the knowledge of this interesting tortoise and promoting captive care strategies that would enable it to live as much as possible according to its biological requirements. It is thus a guideline for the beginner, but it is at the same time intended to convey the fascination these small, active tortoises evoke, their charming personality, the fun of watching them, and how loveable they truly are.

Please enjoy reading this book.

Ricarda Schramm

Introduction

The Egyptian Tortoise, *Testudo kleinmanni*, is the smallest representative of the genus *Testudo* and considered the second smallest recent (presently living) tortoise species after the Speckled Cape Tortoise, *Homopus* (*Chersobius*) *signatus* GMELIN, 1789, which is native to rocky semideserts of western South Africa.

Testudo kleinmanni was first described and so named by LORTET in 1883. It was also the last species of all Mediterranean Tortoises (genus *Testudo*) to receive investigative attention. Even today, much less is known about the ecology of the Egyptian Tortoise compared to the other species of *Testudo*, and new research in its natural distribution range is hampered by the political situation in the countries where it occurs. Its natural range used to extend along the coast of the Mediterranean Sea from northeastern Libya via Egypt to Israel. Within this distribution range, *Testudo kleinmanni* inhabits a narrow, usually 50 to at maximum 90 km-wide strip between the shoreline on the one and desert on the other side. The landscapes here are shaped by a semi-arid to arid (dry) climate, which will create steppes, semideserts and marginal desert zones (Israel) with dry, hot summer months and precipitation limited to the winter, even though the levels of relative humidity remain high throughout the year. This partly rather hostile and unaccommodating environment is the reason for the very typical seasonal activity pattern of the Egyptian Tortoise. It is in fact the only Mediterranean tortoise that will regularly insert a period of aestivation (summer dormancy) as an adaptation to the local climate.

Over the past hundred years or a little more, the natural population of this species has shrunk at an ever-accelerating pace from originally dense, interconnected populations to few individuals in scattered insular areas. By now, the Egyptian Tortoise is considered extremely endangered and immediately threatened with extinction. *Testudo kleinmanni* is therefore currently classified as critically endangered in the Red List of the IUCN (International Union for Conservation of Nature and Natural Resources) and listed in Appendix I of CITES (Convention on International Trade in Endangered Species of Wild Fauna and Flora). In a last-ditch drive to save it, local conservation zones have in recent times been demarcated to which Egyptian Tortoises were re-introduced on the one, and conservation breeding programmes involving private breeders and zoological gardens were initiated in Europe and globally on the other hand.

Perhaps we private keepers and breeders can help this acutely endangered species survive by producing enough offspring to cover the domestic demand and in this manner save uncounted Egyptian Tortoises from being taken from the wild and smuggled to Europe and the USA. This has unfortunately remained common practice until today and therefore still contributes to the decimation of the natural populations.

The present book consolidates as many descriptions, references, husbandry reports etc. as possible with our own observations, experiences, realisations and contemplations. It describes the necessary preconditions, setup and required technology that will enable these animals to live in an environment that will be similar to the situation they find themselves in in nature, including the natural annual climatic cycle. To this end, we portray various options, discuss their advantages and disadvantages, and provide a detailed "annual schedule" that will specify the management of the environmental parameters. Moreover, we offer information on the legal implications as well as recommendations on the acquisition of specimens and how to best adjust them to their new surroundings.

Owing to its ways of life and adaptation to extreme environmental conditions, *Testudo kleinmanni* is widely regarded as difficult and demanding as to its care in captivity and not suited to the beginner keeper. It is still rather rarely kept and propagated in human care. This reputation may be an artefact that has been persisting from the early days, though, when, for one, wild-caught import specimens were kept that tended to find it hard to adjust to new conditions and often perished, and, for the other, knowledge of the way of life of this tortoise and experiences with its husbandry were very limited. With the technological means available today, the much improved state of knowledge of their needs, and maybe even with the aid of the present book, it is no longer much of a problem to keep and propagate Egyptian Tortoises in a naturalistic setup – and they are a source of great enjoyment, too.

Testudo kleinmanni
Photograph Franziska Biedenweg

Systematics

The following brief foray into the systematics and taxonomy demonstrates how difficult it has become to keep up with the constantly changing systematic arrangements and the resultant nomenclature. Species are again and again reviewed and taxonomically re-arranged until today, facilitated by modern methods of genetic research that will provide detailed evidence of how individual taxa are related to each other.

Within the order of the tortoises (Testudinata, Testudines, previously also Chelonia) the Egyptian Tortoise forms part of the suborder of the hidden-necked turtles (Cryptodira), further of the family of the true tortoises (Testudinidae), and finally of the genus *Testudo*.

In English, the genus name *Testudo* is understood as referring to all Palaearctic or European tortoises even though the latter name would by default exclude the Egyptian Tortoise as well as some subspecies of the Spur-thighed Tortoise, *Testudo graeca*. The alternative common name "Mediterranean Tortoises", on the other hand, includes *Testudo kleinmanni*, for this name is based on a wider geographical region and the associated climatic criteria.

The Egyptian Tortoise was first scientifically described by the German scientist and curator at the British Museum, Albert GÜNTHER, in 1869. He named it *Testudo leithii* after Dr. A. H. Leith who had collected the type specimens and made them available to GÜNTHER. As their provenance, GÜNTHER specified Sindh, Baluchistan. This was also the first time an illustration of an Egyptian Tortoise was published.

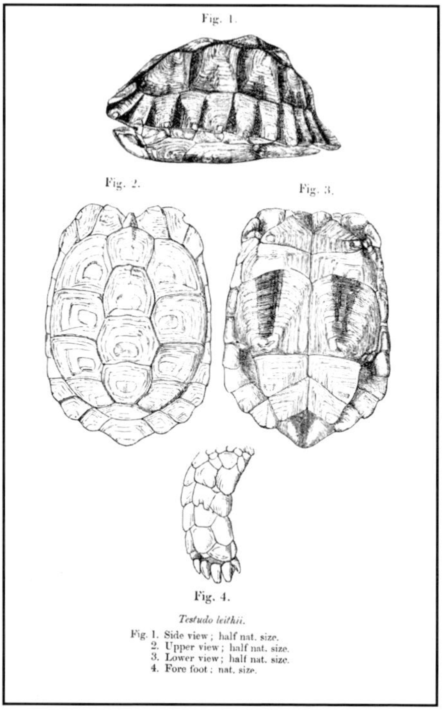

Drawing of Testudo leithii from 1869

Representation by Grönvold in Anderson (1898)

Another illustration was then provided by H. Grönvold in Anderson (1898).

In 1883, Louis-Charles Lortet again described the Egyptian Tortoise, giving it the name *Testudo kleinmanni* in honour of Édouard Kleinmann, a director of the bank Crédit Lyonnais in Egypt and a great lover of tortoises, who had collected a large number of these tortoises in the surroundings of Alexandria and sent them to Lortet for study in 1875. It then took only a few years before Lortet realised that his *Testudo kleinmanni* and Günther's *Testudo leithii* were the same species (Lortet 1887). He voiced doubts as to this species' existence in both Egypt and Pakistan but nowhere in between and suggested that Günther's type specimen actually originated from Egypt as well and might have accidentally been placed in a collection of finds from Sindh. At this stage, the laws of priority determined that the name introduced by Günther was the valid one because it had been published earlier. It then took until 1953 when Ernest Williams, a renowned herpetologist at the Harvard Museum of Comparative Zoology, pointed out that the name *Testudo leithii* had been introduced by Henry John Carter for a fossil turtle already in 1852 (Williams 1953). As a consequence of "*Testudo leithii*" being preoccupied thus, the next available name, i.e., *Testudo kleinmanni* Lortet, 1883, was to be used instead and *Testudo leithii* Günther, 1869, became its senior synonym. Further instances of naming and renaming followed until 1989, including the combination *Pseudotestudo kleinmanni*, but none of these found acceptance.

In 2001, Jarmo Perälä defined another purported species as *Testudo werneri* from the northern Negev Desert in Israel, whose morphological differences he deemed sufficient for recognising it as a separate taxon. Pieh & Philippen (2007) still listed the two species next to each other, but already announced that work in progress was being undertaken to clarify their taxonomic status.

Baha El Din (2006) then was the first to demonstrate that there was only one species, with the same conclusion being arrived at based on genetic evidence by Siroký & Fritz (2007) and Attum et al. (2007).

This means that you should widen your search to the synonymous names *Testudo leithii* and *Testudo werneri* when looking for literature or information on *Testudo kleinmanni*.

***Testudo kleinmanni* in its natural habitat in the Negev Desert of Israel**
Photograph Aviad Bar

Morphology

General Traits of Tortoises

Shell: The probably most conspicuous trait of all chelonians is that their bodies are encased in a shell and this will set them apart from all other vertebrated animals. While the shell will vary in shape, colour pattern and solidity with the species, its basic architecture follows the same pattern in all recent (living today) chelonians. The development of such a shell can rightfully be termed a success of evolution, because what began merely as a thickened layer of skin developed via a string of intermediate stages to today's form of a shell and has been maintained nearly unchanged for the largest part of the evolutionary history of chelonians, i.e., for some 250 million years.

Bone shell with severed bridges and skeleton

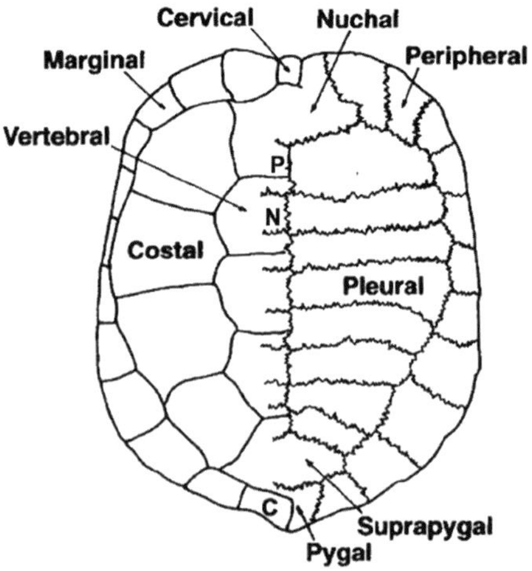

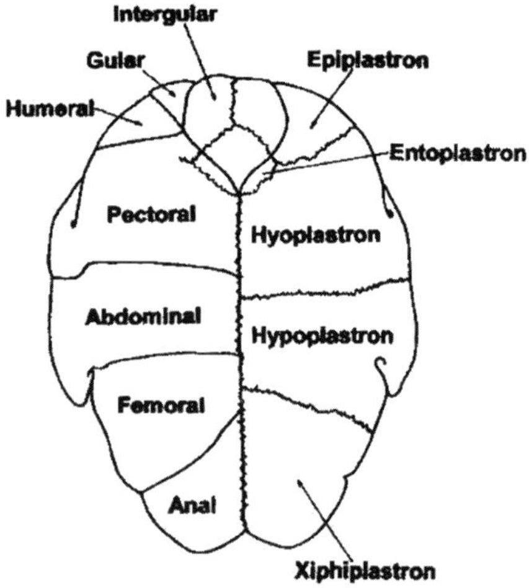

Carapace, dorsal shell
Horn scutes on the left, bone plates on the right

Plastron, ventral shell
Horn scutes on the left, bone plates on the right

Positions and names of the horn scutes and bone plates

Carapace:

Cervical = nape scute; Marginal = margin scute; C: Caudal = tail scute;
Costal = rib scute, also Pleural; Vertebral = spine scute;
Nuchal = nape plate;
P: Preneural = pre-vertebral plate; N: Neural = vertebral plate;
Peripheral = marginal plate; Pleural = rib plate;
Suprapygal = coccyx plate; Pygal = tail plate

Plastron:

Intergular only in Pleurodira (side-necked turtles); Gular = throat scute;
Humeral = arm scute; Pectoral = chest scute; Abdominal = belly scute;
Femoral = leg scute; Anal = cloacal scute
Epiplastron = epiplastral plate; Entoplastron = entoplastral plate;
Hyoplastron = hyoplastral plate; Hypoplastron = hypoplastral plate;
Xiphiplastron = xiphiplastral plate

Bone shell of a Testudo (Agrionemys) horsfieldii, carapace above, plastron below. Clearly discernible are the serrated seams between the bone plates and the outlines of the offset overlaying horn scutes.

Photographs Ricarda Schramm

A tortoise shell is composed of an inner bone layer, the bone shell, and an outer layer of horn scutes, the horn shell. These two layers are separated by a layer of skin with a multitude of nerve cells and a good blood supply to both the bone and the horn layers. It also ensures a firm hold of the horn scutes.

The individual bone plates (osteoderms) have formed by the vertebral column, the broadened ribs, and the pelvic girdle fusing during the course of evolution, resulting in spine, ribs, and parts of the shoulder and pelvic girdles becoming firmly attached parts of the bone plates. The latter are finely interdigitated, which adds to their stability.

The dorsal (carapace) and ventral shells (plastron) are connected on either side via a bone structure known as bridge. This architecture leaves only the cervical and the caudal sections of the vertebral column as well as the front and hind legs freely movable, i.e., all body parts that extend from the openings in the shell in the front and in the rear, respectively, and into which they may be more or less deeply retracted in a case of danger.

This means that the shoulder and pelvic girdles are virtually encased within the chest cavity or rib cage in chelonians, which sets them apart in this aspect from all other vertebrates in which these are positioned outside of the chest cavity.

The numbers and arrangements of the bone plates and the horn scutes above them are not coextensive. Rather will the horn scutes each cover several bone plates and thus also overlap the sutures between them, providing additional strength to the shell. Exceptions exist in species in which some parts of the plastron are mobile; here, the outlines of the bone plates and horn scutes will approximately match each other to facilitate the mobility of the movable segment.

Hatchling tortoises do not initially have an ossified shell substructure, with theirs consisting only of cartilage and connective tissue. It actually takes a few years for the latter to transform into bone matter (ossification), and it may take until their reaching sexual maturity for this process to be completed. Until then, there will be gaps in the bone shell, quite similar in manner to the development of the skull in humans in which these gaps (fontanels) are present in toddlers and will gradually grow closed only as the child grows older.

The tortoise shell grows through the deposition of tissue concentrically around the bone plates and horn scutes, respectively, which is visible in the appearance of growth rings around the horn scutes.

Growth rings around the horn scutes in Testudo hermanni hermanni

Photograph Ricarda Schramm

Limbs: The head, neck, legs and the tail are also armoured with horn in the shape of scales, which may be very small in some areas so that the skin will be soft, and large in others to provide protection.

On the head, sharp-edged horn sheaths (rhamphotecae) on the upper and lower jaws create a beak-like structure that will enable chelonians to bite off pieces even from harder and more solid food items.

Horn sheaths on the upper and lower jaws form a beak:

Testudo hermanni hercegovinensis

Photograph Ricarda Schramm

The feet are furnished with five clawed toes in most species, but these toes may also be reduced in size and/or the claws may be fewer in number in some. In fact, the number of visible claws on the front and hind feet varies between certain species and may even vary within the same species.

The presence, shape and arrangement, and the colouration of certain scales on the limbs and of the horn scutes of the shell are characteristic traits that serve to distinguish between species and subspecies.

Inner organs: The arrangement and functions of the inner organs within the shell are practically identical in all chelonians.

The lungs usually fill the top third of the body cavity inside the shell. Owing to the rigidity of the shell, inspiration (breathing in) is passive in chelonians and expiration (breathing out) is actively achieved by the force of muscles along the surface of the lungs and supported by movement of the front legs and head. With the lungs being positioned above all other inner organs, breathing will be impeded by the weight of the latter when a tortoise is in an upturned position, which may even prove lethal if the animal cannot turn back onto its feet.

In herbivorous ("vegetarian") chelonians, the intestines, and the large intestine in particular, are extremely well developed and long, quite in contrast to omnivorous ("eating everything") and carnivorous ("meat-eating") species. The large intestine is the place where intestinal microbes process the constituents of the ingested plant matter.

Intestine, urethra and oviducts, respectively spermatic ducts, together end in the cloaca.

With the urethra not ending in the urinary bladder, urine may either be discharged via the cloaca or returned to the urinary bladder where the water contained may be recycled by means of resorption. This is a survival mechanism in times of drought.

As end products of the nitrogen (protein) metabolism, chelonians produce ammoniac, urea and uric acid, with the relative proportions of these being correlated to the natural habitats of the individual species. Tortoises, and those from arid (dry) regions in particular, will mainly excrete uric acid and its salts (urates), because this is a particularly water-conservative strategy (uricotelism). Urates are then found as a creamy white component in the urine.

Urates are visible as a creamy white substance next to faecal matter

Photograph Franziska Biedenweg

Sensory organs: Chelonians orientate themselves within their environment with the aid of their sensory organs, eyes, nose, ears, tongue and skin, which also serve to detect predators, food and water, assess the position of the sun, evaluate temperatures, find mating partners, test the moisture content and thus the suitability of potential oviposition spots, and many other things.

The eyes are positioned on the sides of the head and so facilitate a field of vision almost all around. Each eye is outfitted with an upper, a lower and a third eyelid (nictitating membrane, membrana nicitans), as well as two tear glands. The eye is closed by lifting the lower eyelid. The nictitating membrane can be pulled across the eye in an active manner, e.g., for protecting it.

The faculty of vision is largely shifted towards farsightedness, and colour vision is well developed. This enables tortoises to detect their favoured red and orange flowers even from a distance. Additional receptors facilitate colour vision in the UV spectrum.

The ears do not have pinnae (auricles) so that the tympanum (eardrum) is visible as a darkly pigmented, oval spot in the skin surface behind the eye and cheek. Tortoises are able to hear only relatively low frequencies of 100–700 Hz.

The sense of smell is also well developed. Tortoises therefore use the nose to examine in detail potential food, mating partners, and oviposition spots. In a captive setting it may thus be observed that new additions to an existing colony of tortoises will be "sniffed" very thoroughly in the beginning, indicating that established specimens recognise them as "alien" by their individual body scents.

Eye and tympanum (marked),

Testudo hermanni hermanni

Photograph Ricarda Schramm

Nares,

Testudo hermanni hermanni

Photograph Ricarda Schramm

Sexual dimorphism: Sexing by means of external traits will be impossible in juvenile (sexually immature) tortoises for their first few years of life. Reaching sexual maturity is not necessarily a function of age, but rather of having grown to a certain size, and it may vary from one species to the next. The sexual identities of adult tortoises, on the other hand, are quite readily made out on the basis of external morphological markers. In most tortoise species (70 %) the females will be larger than the males. The tail of a male is typically longer than that of a conspecific female and broader at its base. The cloacal opening of a male will furthermore be more distant from the suture of the anal scutes. In some species, the males sport a concave (arched inwards) plastron, which will be of advantage during copulation. Lastly may there be differences in the shapes of the femoral and, especially, the anal scutes between the sexes in some species.

More detailed descriptions of the anatomy of chelonians can be found in the standard reference literature on chelonians (e.g. EGGENSCHWILER 2000, KIRSCHE1998, KÖLLE 2009, ROGNER 2008, SASSENBURG 2005, VETTER 2006).

Sexual dimorphism:
male on the left, female on the right,
Testudo hermanni hermanni

Photographs Ricarda Schramm

Characteristic Traits of *Testudo kleinmanni*

Various authors have described the morphology of the Egyptian Tortoise (ANDERSON 1898, BAHA EL DIN 2006, DELFINO et al. 2009, ERNST & BARBOUR 1989, LAPPARENT DE BROIN et al. 2006a, b, LORTET 1883, 1887, MACALE et al. 2011, SCHLEICH 1989, SCHLEICH et al. 1996, VETTER 2002).

Shell: The carapace (dorsal shell) is oval in outline, very highly domed, and steeply sloped on the sides. A nuchal scute (cervical) is present; it is usually

Highly domed dorsal shell

Photograph Franziska Biedenweg

wide and triangular in shape, rarely long and narrow, and typically clearly projecting. The caudal plate (pygal) is slightly upturned in lateral view and exhibits a concave (inwards) arch and a clearly visible groove on the underside. The caudal scute is elongated beyond the marginals and typically undivided (entire).

The caudal scute is grooved on the underside

Photograph Franziska Biedenweg

Entire caudal scute

Photograph Franziska Biedenweg

The colour of the shell is very light, ranging from uniform yellow to reddish yellow. The horn scutes of the carapace do not exhibit a central pattern, but the growth rings around each of them are initially brown or even black. This colouration is partly lost in older specimens, usually from an age of about five years.

Growth rings around the horn scutes

Photgraph Franziska Biedenweg

The ventral shell (plastron) is noteworthy for its pair of thickened gular scutes that will rarely extend beyond the anterior margin of the plastron. Markings are scarce save for usually a pair of triangular, brown to black elements pointing backwards on the abdominal scutes. Some specimens may furthermore sport another pair of smaller triangles on the pectoral scutes in front of these. These markings may be completely absent, though, leaving the entire plastron uniform light in colour.

The altogether very light colouration protects the Egyptian Tortoise for one against overheating and for the other from ready detection by predators, blending in nicely with the light-coloured sand ground of its habitats. The colours of the shell and skin are in fact almost indistinguishable from that of the sand, concealing these tortoises very well from the eyes of predators.

The light colouration of both shell and skin makes the Egyptian Tortoise blend in with its surroundings, rendering it almost invisible to predators on the sandy ground.

***Testudo kleinmanni* in its natural habitat in the Negev Desert, Israel**

Photograph Aviad Bar

Colouration of carapace and plastron;
The triangular, distinctly protruding nuchal scute can be seen on the left.

Photographs Franziska Biedenweg

A species-indicative trait is the movable xiphiplastron, the bone plate that is covered by the femoral and anal scutes, which can be flipped up or down for a little distance. While this will improve the protection against predators it also enlarges the space available for eggs and makes laying them easier for females.

The bridges between the dorsal and ventral halves of the shell are very wide, and there are two axillar and one or two inguinal scutes.

Another noteworthy trait is that the thickness of the bone shell is greater relative to the shell size than in other species of *Testudo* and this is true for both carapace and plastron (Lortet 1887, Delfino et al. 2009).

Hatchlings will initially sport a deviant colour pattern of the carapace, i.e., a brownish marbled pattern of the scutes, that will fade over the first few weeks to months. The plastron, on the other hand, will already be light in colour, and the dark markings, if at all present, are usually recognisable only as indications.

Juvenile colour pattern, top: carapace, bottom: plastron

Photographs Franziska Biedenweg

Limbs: The rather small to moderately large, narrow head and the limbs are light in colour as well.

A highly characteristic feature are the three longitudinal rows of enlarged scales down the anterior face of the front legs from amongst which the fourth ones towards the body are particularly large and protruding. These serve to strengthen the body armour when the limbs are retracted into the shell.

Enlarged scales on the front legs

Photograph Franziska Biedenweg

The front feet end with five claws whereas the hind feet have four (LOVERIDGE & WILLIAMS 1957), but the presence of five claws on the hind feet has also been mentioned (Schleich 1989). The heels are equipped with enlarged and elongated scales. For their part, the thighs lack the spur-like scales that are so typical of the Spur-thighed Tortoise, *Testudo graeca*.

The tip of the tail usually carries no horny spike, but a very small one may be present in some individuals.

Dimensions: The Egyptian Tortoise is the smallest representative of the genus *Testudo* and the second smallest tortoise species in the world. It grows to an average of 10 cm in carapace length and an average weight of

200–300 g. Males rarely grow larger than 10 cm at weights of 155–200 g, whereas females will typically reach lengths of 11–13 cm at weights between 200 and 300 g (Baha El Din 2006, Schleich et al. 1996). The largest female found as yet measured 14.4 cm (Farkas et al. 1997) and another female with nearly the same carapace length at 14.1 cm tipped the scale at 471 g (Schneider & Schneider 2008).

Life expectancy: The maximum life expectancy was originally thought to be about 26 years in females and about 22 years in males (Macale et al. 2009), but more recent verbal communications now have emerged from keepers who obtained adult specimens as long ago as during the 1990'. They report that these are still alive, evidently rendering them older than that.

Sexual dimorphism: Delfino et al. (2009) and Macale et al. (2011) investigated the sex-related and therefore sex-indicative traits of *Testudo kleinmanni* – Delfino with traditional means in preserved museum specimens and Macale with modern methods of geometric morphometry by measuring shapes of, and distances between, a multitude of measuring points and analysing these data with computer software. They had available 126 live specimens from a shipment of wild-caught specimens illegally imported from Libya and confiscated in Italy.

Like with most tortoises, the females of *Testudo kleinmanni* grow larger and heavier than the males (see above); a larger body volume could have evolutionary advantages with regard to clutch size.

The sexes also differ in the shape of the carapace, which the shell of a female being rather pear-shaped in lateral view because its anterior and posterior curvatures are not quite symmetrical. The carapace of a male, in contrast, appears almost semispherical in its lateral outline. In both sexes, the highest point lies at about the third vertebral scute.

The same perspective furthermore shows that the caudal scute will be more flared in the male than in the female. Also, the groove that is visible on the underside of this scute (see above) will be slightly more distinct in the male.

In dorsal view, the carapace appears narrower in its rear portion in the female and more widened in the male. At the same time, the caudal plate (pygal) and the caudal scute of the male is a little larger, more arched, and projecting outwards. This shape has a facilitating effect on copulation by providing more space for positioning the long tail (Macale 2011).

Body shape in lateral view,
top: male, bottom: female

Photographs Franziska Biedenweg

Body shape in dorsal view,
top: male, bottom: female

Photographs Franziska Biedenweg

In ventral aspect, the plastron of the female tends to be elongated at the midline and a little widened at the levels of the pectorals and abdominals. The posterior portion of the plastron narrows towards the anal scutes.

In contrast to the widening in the rear portion of the carapace, the plastron of a male narrows substantially towards the rear so that the openings for the hind legs will be altogether larger than in females. This could help the male to move with greater agility and swiftness than a female and so be of advantage in courtship. A wider rear body could also provide more freedom of movement to the legs and more stability during male/male combats.

The plastron of male *Testudo kleimanni* is flat and not concave as in many other species of *Testudo*.

The shape of the xiphiplastron (or the femoral and anal scutes) differs between the sexes in that it will be wider in its posterior portion in the male than in the female. This is a result of the xiphiplastron's greater dimension in the region of the midline in males, that is in the border area between the femoral and anal scutes, and a longer middle suture between the anal scutes (DELFINO 2009)); if you draw a line in this area, you can actually measure this difference. The angle formed by the sutures between femoral and anal sutures is often larger in male specimens, and these sutures may sometimes be curved towards the tail. This angle is more acute in females, and the sutures will be rather straight. A direct comparison of this trait is of course possible only in specimens of about the same size. Another difference between the sexes must not be omitted here: The shape of the suture between the abdominal and femoral scutes is slightly different. In a male, it will be more curved towards the anal scutes in the middle and more curved towards the abdominal scutes on both sides. In contrast, this suture will be more or less straight in a female. This difference is likewise most clearly recognisable by drawing a line as suggested by the drawings published by DELFINO et al. (2009).

With all these comparisons one should keep in mind that nature does not always follow fixed templates and that these traits may therefore not be expressed equally clearly in all individuals of a species. Exceptions from the rule must be expected at all times.

Body shape in ventral view,

top: male, bottom: female;

visible here as well is the movability of the slightly opened xiphiplastron

Photographs Franziska Biedenweg

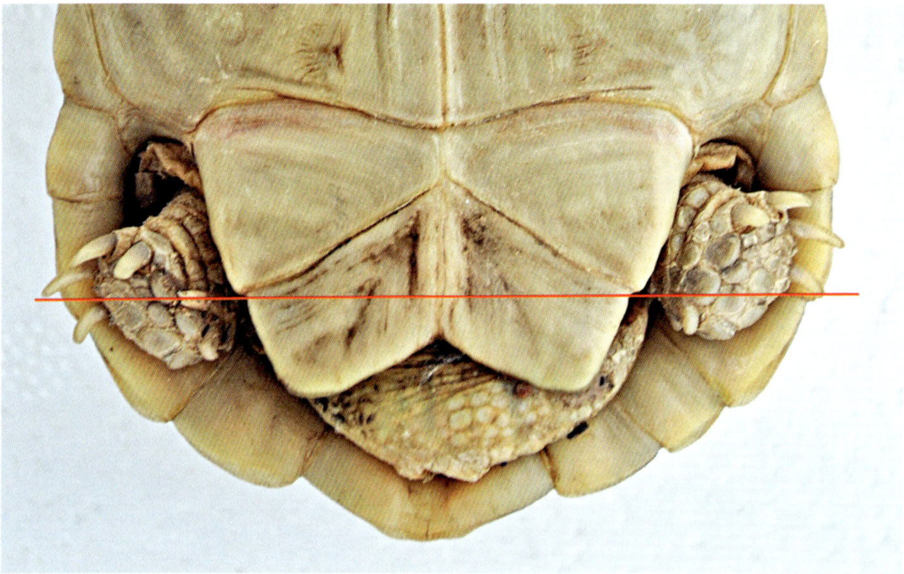

Differences in size and shape of the anal scutes in the male (top) and in the female (bottom)

Photographs Franziska Biedenweg

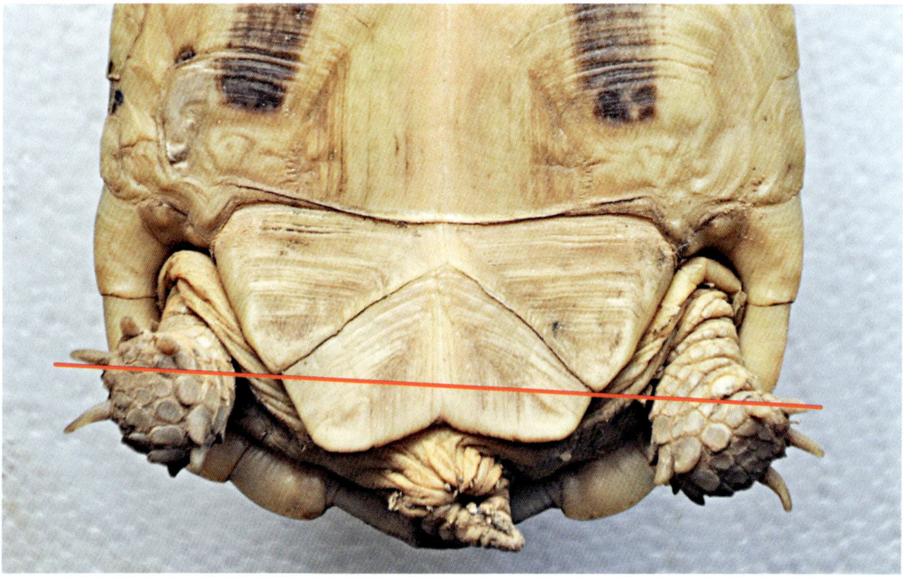

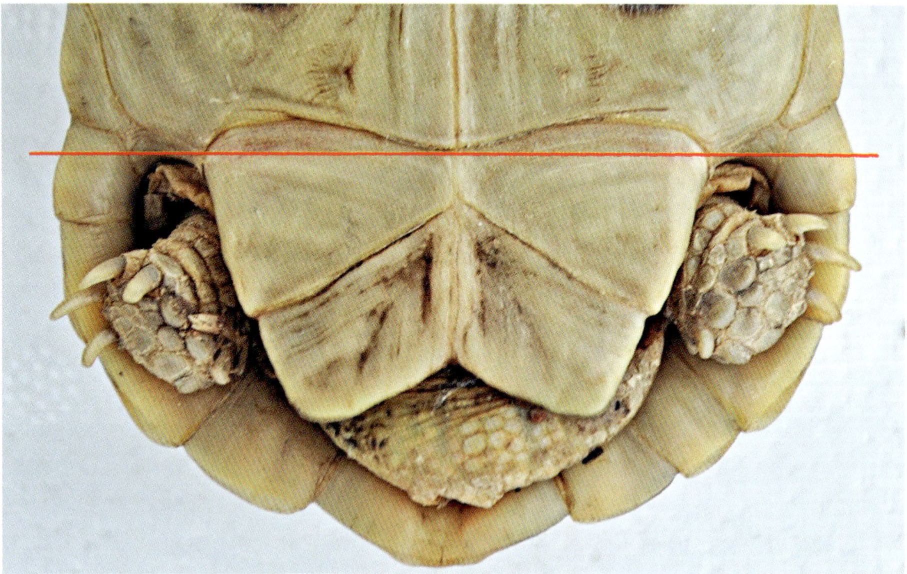

Differences in the suture between the abdominal and femoral scutes, top: male, bottom: female

Photographs Franziska Biedenweg

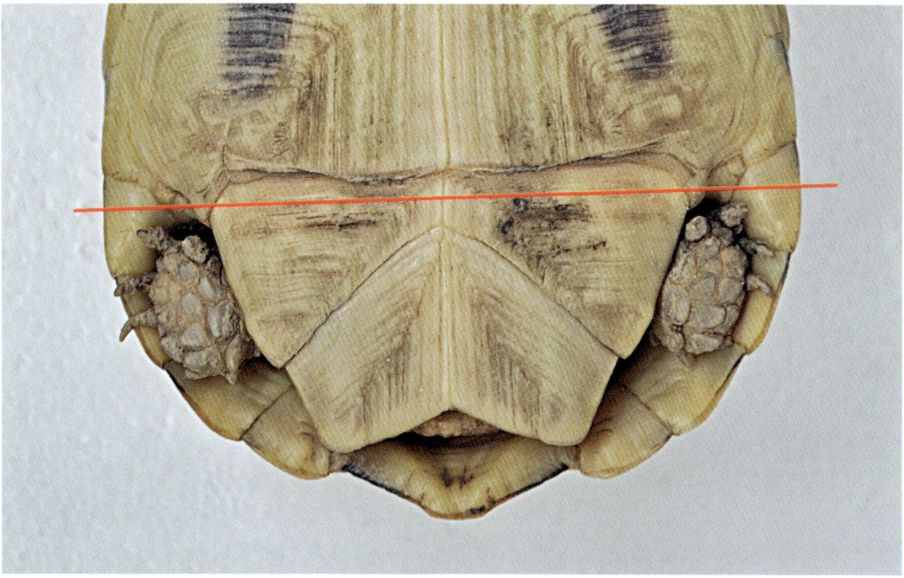

Natural Habitats

When keeping tortoises in human care it is of utmost importance to create conditions that will match those the animals live under in the wild as closely as possible. "As closely as possible" must here make provision for the fact that it will always be possible only to a limited extent to emulate with technological means the conditions in nature. And to achieve this, it is an absolute necessity to know the climate, soil, vegetation that all make up natural habitats together and influence the ecology of the tortoises that live in them.

Distribution

Around the end of the 19th century, the Egyptian Tortoise was still so common in the country to which its vernacular name refers that the banker Edouard Kleinmann did not have much trouble to collect a larger number of specimens for shipping them to LORTET (1883). It is likely that a more or less blanketing distribution range still existed at this time, extending from northwestern Libya (Tripolitania) via the coastal strip along the Gulf of Sirte (Sirtica) to northeastern Libya (Cyrenaica or Barka) and on to northern Egypt and the Sinai Peninsula to the northern parts of the Negev Desert in Israel.

Libya

First records of the existence of *Testudo kleinmanni* in Libya surfaced in the early 20th century, in particular from the northern Cyrenaica, near the town of Darna in the northern coastal strip and the northeast of the country, some 40–80 km inland on the border to Egypt. These finds were reported upon by WERNER (fide SCHNEIDER & SCHNEIDER 2008). Dr. Henry W. Setzer then collected three specimens in the vicinity of the town of Qaminis in the summer of 1955 that he donated to the United States National Museum, today's National Museum of Natural History, Smithsonian Institution (BUSKIRK 1985). LOVERIDGE & WILLIAMS were the first to mention its occurrence in the surroundings of the town of Sirte (1957). In the framework of his herpetological fieldwork in northeastern Libya, Hans-Hermann SCHLEICH investigated the distribution and biology of the three tortoise species native to the Cyrenaica, including the Egyptian Tortoise. He located populations in five places within the northeastern portion of this region, including the El-Kouf National Park south of the Jabal al Akhdar, which used to be known as the Barka Plateau in earlier times (SCHLEICH 1984, 1987, 1989, SCHLEICH et al. 1996). During the course of the 1980', Zych collected a few specimens on the northern slopes of the Jabal Nafusah and about 50–60 km east of Tripoli near the coast (SIROKÝ & FRITZ 2007). FRITZ & BUSKIRK (1997) then were the first to specify localities at which *Testudo kleinmanni* occurred in Tripolitania, i.e. between Tarhunah and Bin Gashir as well as at Al Khums.

The beginning of the 21st century saw several more authors reporting on the results of their expeditions and field studies, confirming occurrences of *Testudo kleinmanni* in the known regions. FRYNTA et al. (2000) for example, mentioned a single dead specimen that was found at the archaeological dig site of Leptis Magna in the vicinity of Al Khums. For his part, BAHA EL DIN (2002) located this species to the east and also 30 km to the west of Tobruk and on the southern fringes of the Jabal al Akhdar in the surroundings of the town of Mekhili. Not far from here, PIEH likewise found specimens of *Testudo kleinmanni* on the road to Tobruk, 20 km south of Derna as well as in Wadi Derna and at Leptis Magna (PERÄLÄ 2005). Christoph and Willi SCHNEIDER travelled through Libya in 2005, 2006 and 2007 and documented their field studies in great detail: Their route took them from the Tunisian border east along the Libyan coast to the harbour town of Tobruk, and they recorded populations from all three Libyan provinces in the process. Like FRYNTA had done five years earlier, they found live specimens of *Testudo kleinmanni* in the surroundings of the digging sites of Leptis Magna. These authors furthermore again recorded the Egyptian Tortoise from the province of Sirtica and thus confirmed the record by LOVERIDGE & WILLIAMS (1957) from the town of Sirte half a century earlier. In their first report on their travels in Libya in 2005, they discussed the purported occurrence of *Testudo kleinmanni* at Suluntah, as had been mentioned by PERÄLÄ (2005) and SCHLEICH (1989)), and concluded that its existence here was unlikely due to the given habitat conditions. In total, they documented 28 locality records from along the coastal strip across a maximum width of 90 km inland from the coastline and in the Cyrenaica (SCHNEIDER & SCHNEIDER 2006, 2008).

When Wolfgang HLAWATSCH spent time on business in Misurata between 2006 and 2008, he happened upon large numbers of *Testudo kleinmanni* that were offered for sale in local pet shops. Wondering where these might come from, he set out for himself and eventually managed with the help of some native fellow workers to find one specimen some 50 km to the southwest of Misurata in October of 2008. Local people then confirmed that these tortoises were abundant here in March when everything was green and flowering and could be found in their numbers in the morning, which was when they were collected, even by children, and sold to pet shops (HLAWATSCH 2009).

Egypt and Israel

Most historical records of *Testudo kleinmanni* refer to Egypt (Buskirk 1985).

Lortet (1883) described the type locality in the vicinity of Alexandria as follows (in transl.): "It is found in large numbers in the sandy dunes of Egypt, in particular in the surroundings of Alexandria. It was sent to me in immense numbers by Mr. Kleinmann. It is found from Egypt to as far as Syria, between Ismailia and El Arish throughout the entire region that is known by the name of 'Little Desert'. I have furthermore encountered it in the surroundings of Hebron and in Beersheba on the northern limits of the Desert of Sinai."(Lortet 1883, 1887). "Desert of Sinai" here probably refers to the Negev Desert, for Beersheba (= Be'er Sheva) is situated on the edge of the Negev Desert in southern Israel. This locality was also confirmed at around this time by Tristram (1884).

In 1963, the little tortoise was rediscovered in Israel by Yehudah Werner who conveyed his discovery in a personal communication to Bringsøe and Buskirk, specifying the locality of his find as Nahal Secher, 14 km south of Beersheba (Bringsøe & Buskirk 1998, Buskirk 1990). These latter authors then furthermore mentioned their own finds at Beer Mashabim, 20 km to the northwest of Sede Boker, and at 6 km northeast of Nitzana, and discussed known localities at Holot Haluza, Mishor Yamin, and Mishor Rotem (Arbel 1990, Bringsøe & Buskirk 1998, Mendelssohn & Geffen 1989, 1991, Zinner 1974). *Testudo kleinmanni* thus became known from a handful of localities in the northern parts of the Negev Desert of which the largest and best researched population is the one from Holot-Agur, a protectorate of some 400 km² in dimension between Beersheba and the Egyptian border. Geffen & Mendelssohn (1988, 1989, 1991) then undertook three subsequent journeys to this region in order to investigate various ecological aspects such as the thermoregulation, home range, and reproductive behaviour of the Egyptian Tortoise in a study area of 3 km². 1994 saw Sherif Baha El Din conduct a field study with the aim of assessing the status of this species on the north coast of Egypt and in the northern sectors of the Sinai Peninsula and, after failing to find even a single specimen during his journey, arrived at the conclusion that the Egyptian Tortoise had apparently ceased to exist in its original distribution range in Egypt. The only indication that a very small, isolated population might still persist was based on the statement by a local shepherd who claimed having seen two specimens ca. 32 km east of Salum on the north coast of Egypt in spring. A very similar picture emerged on the Sinai Peninsula where some residents of the region around El Ogra on the Egypt-Israel border near Holot

Agur confidently stated that this was the only area in which tortoises persisted. Farther to the west, BAHA EL DIN met local residents who reported having seen a very few tortoises on the sandbanks and smaller islets on the southeastern shores of Lake Bardawil. Even though he then failed to find one of these specimens, BAHA EL DIN thought it very likely that some individuals or very small and isolated populations might indeed have survived in these situations. He supported this assessment by referring to an older find on the island of Um El Rumiat in the western part of Lake Bardawil in 1985 and the fact that soldiers of the coast guard stationed there had admitted to collecting specimens there. All remnant populations appeared to have finally disappeared by the time he revisited the region in 1990 and found merely a single individual at El Teloul.

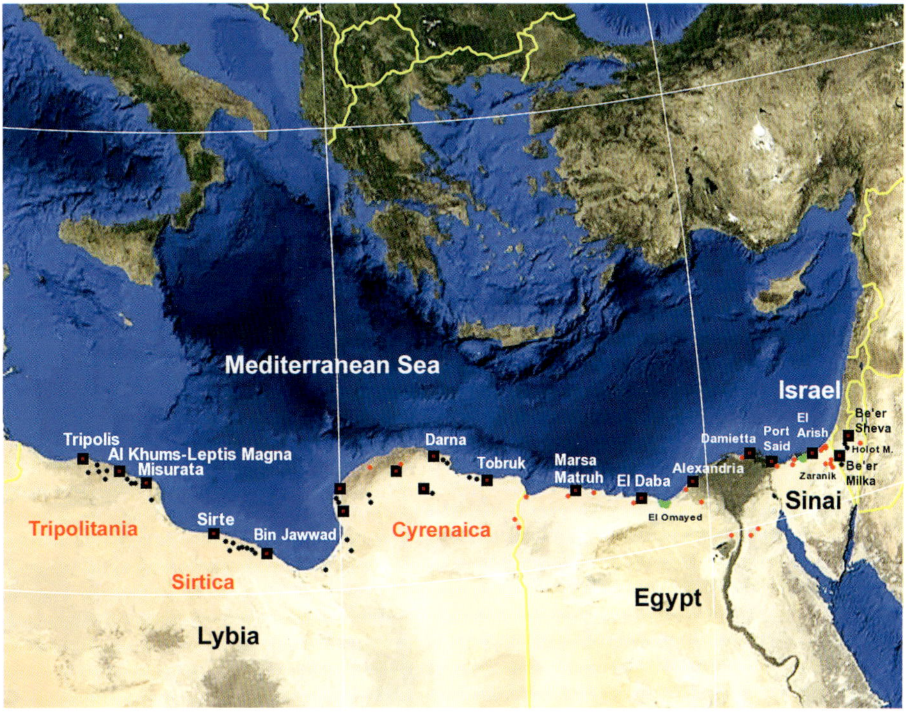

Distribution Map of Testudo kleinmanni

● ***Extinct Populations*** ● ***Persisting Populations***

● ***Conservation areas, reintroduction projects***

Modified ©Franziska Biedenweg

Baha El Din (1994) visited and searched a total of 49 possible localities based on the earlier finds of previous authors and the results of interviewing local residents. He furthermore visited animal trade markets in a quest to assess the extent of trade in tortoises. Earlier records of populations in northern Egypt, on the Sinai Peninsula, and in Israel were published by Flower (1933), Loveridge & Williams (1957) and Mendelssohn (1982).

More recently, Shelley J. Rank & Liz Kaufman (2012) have been monitoring specimens of *Testudo kleinmanni* (referred to as *Testudo werneri*) in the framework of a study on its natural dietary spectrum in the Mashabim Dunes of Holot Mashabim, a nature reserve some 35 km south of Beersheba, i.e. in localities that had already been described earlier by Henrik Bringsøe & James R. Buskirk (1998).

Threats and Conservation

It took just a little more than a hundred years since its discovery for the situation of *Testudo kleinmanni* to turn from very good to very bad in its native countries.

Status in Egypt: Sherif und Mindy Baha El Din reported already in 1994 that the largest part of the former natural habitats along the north coast of Egypt had been nearly completely destroyed and did no longer support the existence of *Testudo kleinmanni*; the same situation was found in the northern parts of the Sinai Peninsula. By now, the little tortoise has vanished practically entirely from its original natural habitats in Egypt save for two or three, locally confined, very small populations (Sherif & Mindy Baha El Din 1994, Baha El Din et al. 2003, Baha El Din 2006). Ramon Mascort (2009) found not a single Egyptian Tortoise on the coast between Alexandria and the Libyan border in 2008. Reasons for this lamentable development here include the destruction of habitats through the expansion of tourist infrastructure, intensified agricultural practices, utilisation of areas for military purposes, but also decades of intensive collection of specimens for the commercial trade. Even though *Testudo kleinmanni* was afforded the status of a protected species as per National Law No. 192 in Egypt already in 1983, animal dealers in Cairo claim to have never heard of it (Deveaux 1997, Sherif & Mindy Baha El Din 1994).

Status in Israel: While collection of, and commercial trade in *Testudo kleinmanni* do not play major roles in Israel, its habitats have continually been destroyed here as well as a result of more intensified land use and military training upon which was reported already in the 1980' (Buskirk 1985).

The Egyptian Tortoise, Testudo kleinmanni
Top: In the Negev Desert, Israel   Photograph Aviad Bar
Bottom: In the conservation area of Omayed (Omayed Biosphere Reserve), Egypt.

Egyptian Tortoises, confiscated in Genoa harbour, Italy

Status in Libya: Considering all the aforesaid, the only still sizable populations of the Egyptian Tortoise left are those in Libya. Here, destruction of habitats play less of a role. Between the major cities of Tripoli in the west and Benghazi in the east, a stretch of nearly 1,000 km of almost primary coastal landscape exists where tourism does not play a role and which is utilised only by local people for grazing their livestock (Schneider & Schneider 2006). Collecting tortoises from the wild continues also in Libya, however, to satisfy the demand by pet shops in Tripoli, by Egyptian migrant workers as something to take home and then smuggle via Tunisia, Malta or Italy for sale on the markets in Europe or the USA (Schneider & Schneider 2008).

In 2006, 400 *Testudo kleinmanni* were seized at Milan Airport when a Libyan national from Tripoli tried to smuggle them into Italy (Philippen 2006), and customs in Genoa harbour also found and confiscated a large number of these tortoises from North Africa. The unfortunate fact is that this smuggling has not come to an end until now, as is evident from confiscations that were effected in Libya in 2017.

Members of The Libyan Wildlife Trust stopped a truck with a cargo of about 3,000 illegally collected tortoises in November of 2017 that were en route to Egypt.

Photographs The Libyan Wildlife Trust

Testudo kleinmanni and *Testudo graeca cyrenaica* were discovered hidden beneath bales of straw.

Photographs The Libyan Wildlife Trust

On 28 April 2017, the Libya Herald, The Independant Libya Online Daily, reported that almost 1,900 Egyptian Tortoises had been seized by Immigration Police in the surroundings of Sidi Khalifa near Benghazi. They were meant to be smuggled to Egypt for further sale (Muftah 2017).

Members of The Libyan Wildlife Trust, a Non-Governmental Organisation (NGO) that was founded in 2011 and partners with the WWF, reported that they were able to stop a truck transporting ca. 3,000 tortoises on 25 November 2017, with the shipment including not only *Testudo kleinmanni*, but also *Testudo graeca cyrenaica*. The animals had been hidden under bales of straw and were likewise to be smuggled to Egypt.

Few of the Egyptian Tortoises imported into Europe and the US in the 1980' and 1990', as well as those smuggled today, survived the stress of excessively long transport routes. If they somehow reach their destinations alive, they will usually be very weak, severely dehydrated, and fraught with huge amounts of parasites (Hobbs & May 1993, Zwartepoorte 2007).

Perälä (2002) estimated the total population of Testudo kleinmanni to have originally comprised 55,624 individuals, which thus exceeded the estimate to the same effect by Mendelssohn (1982), whereas their number today would total a mere 7,470 individuals of which some 75% may be sexually mature. According to this calculation, the total population of Egyptian Tortoises would have declined by 86.6% within less than three (tortoise) generations. The inhabited area was estimated by the same author to have shrunk from ca. 123,600 km² originally to ca. 16,600 km² at present (Perälä 2001, 2002, 2003, 2005, 2006).

Assessment by the IUCN: Amongst other things, the IUCN (International Union for Conservation of Nature and Natural Resources) compiles the red lists of endangered species and categorises conservation areas via the World Commission on Protected Areas. As a consequence of the estimates and reports that were supplied to the IUCN by Perälä and other researchers over several years, die Egyptian Tortoise was classified as "vulnerable" already in 1988, reclassified as "endangered" as from 1996, and reclassified yet again as "critically endangered" from 2003 by the IUCN and incorporated in their Red Lists. The explanation for last decision was that it was to be feared that this species could be completely extirpated within less than twenty years or one (tortoise) generation if the destruction of its habitats and the illegal trade in it could not be stopped.

Assessment by CITES: CITES is short for "Convention on International Trade in Endangered Species of Wild Fauna and Flora" and the result of a resolution by the IUCN member states adopted in 1963 that eventually became effective in 1975. Here, *Testudo kleinmanni* has been featuring on Appendix I since 1994 (Pritchard 1995). This Appendix outlaws, amongst others, the commercial trade in specimens that represent a particularly endangered species and have been removed from the wild. The enforcement of this convention and the trade in captive-bred individuals is taken care of by the individual signatory states.

Conservation projects: Mindy and Sherif Baha El Din in 1994 founded "TortoiseCare Egypt" with the aim of establishing an on-site captive breeding programme for *Testudo kleinmanni* in an area where this species used to occur naturally in the past. They selected for this purpose the Zaranik Conservation Area in northern Sinai that had been declared a national park in 1985 and populated it with an initial founder stock of 200 *Testudo kleinmanni* from confiscations.

Reintroduction programmes have also been initiated in the roughly 700 km² El Omayed Biosphere Reserve (OBR) some 80 km km to the west of Alexandria. After decades of absence, a wild Egyptian Tortoise has been found here, giving hope that a new population might be forming here (Shaltout 2004).

Rescue projects now focus on both free-ranging tortoises right in their natural habitats (in situ) and individuals kept captive in fenced-in areas (ex situ).

The Omayed Biosphere Reserve, Egypt, 2017

Testudo kleinmanni, Omayed Biosphere Reserve, Egypt, 2017

These efforts try to involve the local human populace – because "You can love only that which you know, and you will protect only that which you love". With the trade in Egyptian Tortoises being an everyday occurrence in Egypt and a source of income for people, they just have to be incorporated into the conservation of, and field observation work on, these animals and even be paid for their contributions. To this end, local assistants are first trained through working with a captive population before they are sent out to use their tracking skills and familiarity with the local environment to gather information on the activity patterns and feeding habits of the tortoises. This should lead to an increased level of interest in the conservation of this species, participants can be justifiably proud of their achievements, and it is a source of income. It is an approach that has already proven successful in other African national parks where local people have been involved in the monitoring of poaching activities. At the same time, monitoring studies based on radio-telemetry are conducted. There are two possible options of reintroduction that are investigated and compared as to their effectiveness: the "hard release" that is not preceded by a period of adjustment and the "soft release" that will see the tortoises being relocated during their annual period of aestivation, granting them time to acclimatise while they are inactive. Trials to this effect have demonstrated that the “soft release” method has fewer tortoises leave the area that is to be repopulated, increasing the success of a reintroduction initiative (Attum 2007, Attum et al. 2007, 2008, 2010, Attum & Cutshall 2015, Attum & Rabia 2016).

In Libya, nature conservation authorities and local institutions have indicated interest in conservation and reinforcement breeding projects for tortoises and a possible cooperation with the Egyptian conservation project "TortoiseCare Egypt" (Baha el Din 2002).

In the Holot Mashabim Nature Reserve in Israel, populations of *Testudo kleinmanni* persist in the dunes, but the area is open to the public and used by both hikers and bikers, and drug and arms dealers as well as human traffickers, in addition to the grazing livestock of Bedouins, with all of these impacting on the tortoises living here (Rank & Kaufman 2012).

It can thus only be hoped that the mentioned conservation measures will be expanded further and become successful in the end, for they are the only way to prevent to total extirpation of the Egyptian Tortoise in its native countries.

Captive breeding programmes in Europe: 1997 saw the first stud book for *Testudo kleinmanni* being established in The Netherlands. Initially limited in range to national level, other national stud books subsequently joined in, leading to the founding of the European Studbook Foundation (ESF) in 2003. By 2007, 24 private participants from four European countries were contributing, with their number rising to 45 private keepers from twelve countries by 2015.

In 2001, the European Association of Zoos and Aquariums (EAZA) initiated a European Endangered Species Programme (EEP), which the Rotterdam Zoo has been coordinating with the ESF since 2002.

The two stud books, ESF and EEP, were managed by Henk Zwartepoorte from 2003 to his death in 2016. His goal was to interlink the captive breeding programmes with the conservation efforts in North Africa, as this was deemed the most promising approach to save and make the populations of *Testudo kleinmanni* in Egypt recover (Pieh & Philippen 2007, Zwartepoorte 2004a, b, 2007a, b, 2013, 2015a, b, Zwartepoorte & Verhoeks 2014).

Climate

The Egyptian Tortoise inhabits a relatively narrow strip of land between the coast of the Mediterranean Sea on the one and desert on the other side in North Africa. The maximum distance of locality records of *Testudo kleinmanni* from the sea is 90 km (BUSKIRK 1990, SCHNEIDER & SCHNEIDER 2008). The climate within this ribbon of land is semidesert in character, marked by dry, hot summers with cool nights, and rains in winter when temperatures will be moderate during the day and cold at night, even though values do not decrease below freezing point. The ground is sandy to rocky, heats up to very high values during the days due to the normally minimal cloud cover, but will be unable to store much heat for release at night. This results in partly substantial fluctuations between day and night temperatures.

In winter, the average temperatures reach a minimum of 9 °C at night and a maximum of 17.5 °C during the day in January, which is the altogether coldest month. Temperatures will rise during the course of the year, reaching the highest average values in August with a minimum of 22.5 and a maximum of 32 °C. Thereafter, they will gradually decrease once more, but still be 18.5 °C at minimum and 27.5 °C at maximum on average in October, before closing in on the lowest temperatures in January. Temperature fluctuation of 10–15 °C within 24 hours are common.

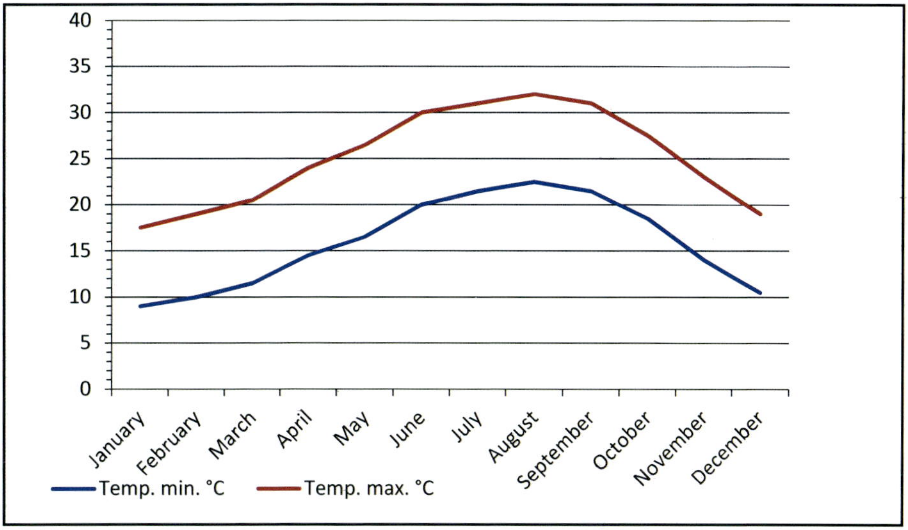

Average maximum and minimum temperatures during the course of the year at Misurata, Libya; data are average values from long-term recording.

©Ricarda Schramm

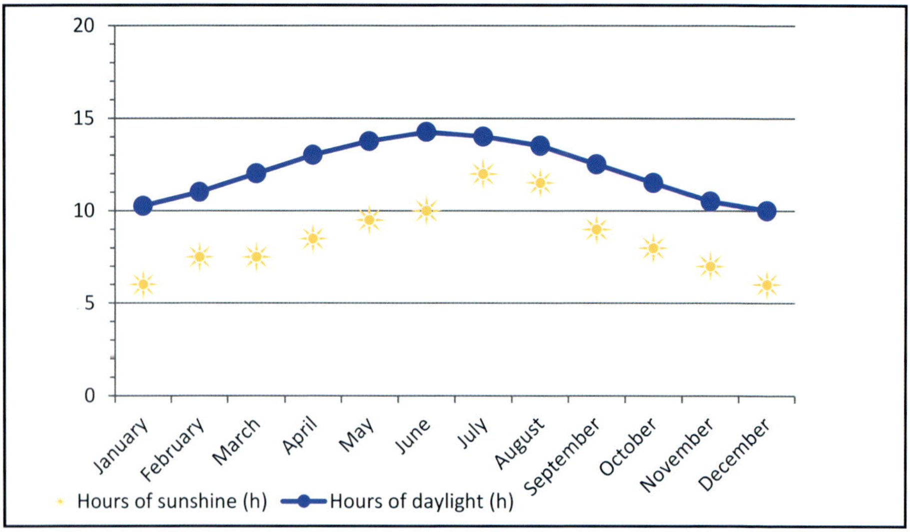

Average hours of sunshine and hours of daylight during the course of the year at Misurata, Libya; data are average values from long-term recording.

©Ricarda Schramm

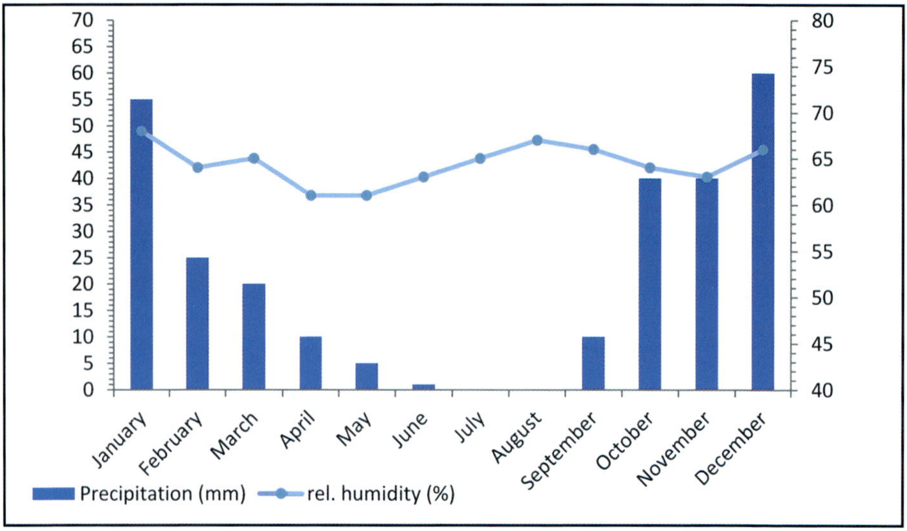

Average precipitation and relative humidity during the course of the year at Misurata, Libya; data are average values from long-term recording.

©Ricarda Schramm

The climatic data reproduced here are average values from ten to twenty years of recorded weather conditions and as such will not provide an idea of the extreme values that are of course experienced every now and then. Attum et al. (2013), for example, measured individual extremes of 54.5 °C during the day and 3.0 °C at night in the Sinai of Egypt during the summer months.

The average number of hours of daylight increases from a low of 10.25 hours in January to a peak of 14.25 hours in June and July. The formation of clouds and rainy weather reduces the hours of sunshine to about 6 during the winter month of January, whereas 12 hours of sunshine are commonly experienced in July and August. This illustrates that the number of hours of sunshine does by no means equal the number of hours of daylight, which is a factor that must be taken into consideration for the husbandry of Egyptian Tortoises in human care. The chapter *Captive Care* will discuss this detail more in depth (p. 65).

Reasonably prolific precipitation is seen only during the winter months, approximately from October through March, with a maximum of 55–68 mm on average in December and January. By May, this average will have decreased to just 5 mm gemessen, and no rain can normally be expected in June, July and August. The rains will resume only in September with an average yield of 11 mm.

Considering the average values of relative humidity in conjunction with those of precipitation during the course of the year, it becomes clear that the former do not form a pattern that would correspond to that of the latter – in fact, the average values of relative humidity are subject to very little annual variation. They range from 61–68 % throughout the year, and still vary only from 63–67 % even in the summer months of June, July and August when no rains at all are experienced. Individual daily values may of course lie significantly above or below the averages. Peak values of between 80 and 90 %, on the other hand, are reached in the winter months at night, decreasing to 50–60 % during the subsequent days. This is a function of the major temperature fluctuations between day and night that will commonly be seen in semideserts and deserts, creating a thawing point near the ground. The latter facilitates prolific amounts of dew to precipitate at night, which in effect forms the basis of life for plants and animals in this rather hostile environment during this period. These conditions are of vital importance for the keeping of the Egyptian Tortoise in human care and mean that corresponding levels of relative humidity must be created all year round, including during the summer months while the tortoises will observe their annual period of aestivation (summer dormancy).The climatic charts reproduced here provide data from Misurata (aka. Misrata), 210 km to the east of Tripoli, Libya. Comparisons of these with those from Sirte (aka. Syrte or Surt), Libya, or Beersheba (aka. Be'er Sheva), Israel, differ only to a minor extent.

Habitats and Ecology

Owing to the troubled political situations in the native countries of the Egyptian Tortoise and the consequently limited opportunities for field studies, reports of more recent date on the natural habitats and ecology of these tortoises are very scant.

The behavioural expressions and the annual life cycle of *Testudo kleinmanni* are responses to the structuring of their habitats and the climatic conditions. Their habitats are confined to a narrow strip of land on the coast, with the maximum distance from the sea shore being determined by the fact that precipitation will greatly decrease the farther inland one travels, reducing the availability of food.

In the west of the total distribution range, in western Libya, the ground consists of compacted sand, partly heaped to form sand dunes in places that are sparsely vegetated with shrubs. Farther to the east, east of Benghazi, the sandy ground gradually transforms into steppe and eventually to rock-strewn and even very rocky areas in the Libyan east (Schleich 1989, Schneider & Schneider 2006, 2008). A steppe landscape is also encountered from the Sinai Peninsula, where *Testudo kleinmanni* used to occur, to Israel, which will likewise transform into sandy ground and sand dunes in the marginal zones of the Negev Desert in Israel (Geffen & Mendelssohn 1988, Mendelssohn & Geffen 1995). The local plant society here is called *Artemisia* steppe or, more commonly, sagebrush. It is a treeless steppe with grasses and bushes, named after the predominant plant species, *Artemisia monosperma*, commonly known as sagebrush, which is a shrub that will grow to a height of 120 cm and more and is related to our native mugwort, *Artemisia vulgaris*.

In areas characterised by sand dunes, the sand is held together by vegetation in which *Artemisia monosperma*, *Ephedra* sp. (a conifer-like herb also known as Mormon tea or, falsely, sea grape), and *Atriplex* sp. (the genus of the oraches) dominate. The sparse vegetation cover of their natural habitats offers Egyptian Tortoises very little in terms of hiding places, covering only about ca. 20–30% of the surface (Geffen & Mendelssohn 1988). Very few reports actually provide lists of the plant species encountered here (Attum et al. 2007, Geffen & Mendelssohn 1988, Gruber 1998, Ibrahim 2001, Schneider & Schneider 2008, Rank & Kaufman 2012).

Artemisia steppe, Sagebrush

A Testudo kleinmanni in its shelter; Lehavim, northern Negev Desert

In order to provide an impression of the plant society, the tables on the following pages illustrate some examples of shots from natural habitats of the Egyptian Tortoise. Plants marked with "S" are those the tortoises use for sheltering, and feeder plants are marked with "F". Not all of the latter are as yet certain to be fed upon by *Testudo kleinmanni*, because no research has so far been undertaken, by faecal examination or stomach flushing, for example, to identify the effectively ingested plants. The classification employed here is merely based upon field observations of the tortoises and forensically interpreting traces left on the plants. Owing to their small size, Egyptian Tortoises can access only low feeder plants.

Species	Family	Photograph
Astragalus caprinus Beer Sheva milk-vetch **S F**	*Fabaceae* Legumes	
Atriplex spec. Oraches **S**	*Amaranthaceae* Amaranths	
Coridothymus (syn. Thymus) capitatus, Mediterranean wild thyme **S**	*Lamiaceae* Mints	
Echiochilon fruticosum **F**	*Boraginaceae* Borages	
Erodium ciconium Common stork's-bill **F**	*Geraniaceae* Geraniums	
Gymnocarpos decandrus **S**	*Caryophyllaceae* Pinks	

Species	Family	Photograph
Malva sylvestris Common Mallow **F**	*Malvaceae* Mallows	
Mesembryanthemum crystallinum Common iceplant **F**	*Aizoaceae* Stone Plants	
Neurada procumbens Creeping-thorn rose **F**	*Neuradaceae* Thorn Roses	
Plantago albicans Downy plantain **F**	*Plantaginaceae* Plantains	
Retama raetam Retem **S**	*Fabaceae* Legumes	
Thymelaea hirsuta Shaggy sparrow-wort **S**	*Thymelaeaceae* Daphnes	

Shelley J. Rank and Liz Kaufman (2012) examined the active agents of a selection of feeder plants and in part obtained surprising results. Protein contents averaged 17.3 %, and the content of raw fibre ranged from 33.4–49 %, which were values right within the range of what tortoises should be fed with in human care. To put this into perspective: The protein content of dandelion leaves is about 17.8 %, and red clover comes to 18.5 %, but lettuces have protein contents of more than 25 % and commercial feeds (sticks) even of more than 30 %. Examples for raw fibre contents are red clover 22.0 %, curled mallow 24.7 %, meadow hay 31.0 %, but iceberg lettuce only 11.1 % (Dennert 2008). Carolin Dennert recommended as a benchmark value for protein contents an upper limit of 20 % and a minimum content of at least 12 %, better 20–30 % for raw fibre. The surprising part of Rank & Kaufman´s (2012) study were the calcium-phosphorus ratios (Ca:P) of the plants they analysed. They had expected values of about 5.1 to 8.1, but found the ratio to be in fact around 14, on average, which is a rather high value indeed. This becomes even clearer when compared to data published by Carolin Dennert (2008) for some native plants, such as broadleaf plantain at 6.1, dandelion at 2.3, red clover at 4.7, but iceberg lettuce only at 0.9, and common kitchen lettuce at a mere 0.02. Dennert recommended a Ca:P ratio of at least 1:1, better 1.5:1 to 2:1, i.e., one and half to two times as much calcium as phosphorus. Another surprise was the results of their analysing the contents of oxalic acid that produced values of more than 0.50 g pro 100 g in all plants examined. Normally, an oxalic acid content of more than 0.30 g per 100 g is deemed too high in tortoise feeder plants (e.g., Senneke 2004), because studies in mammals have demonstrated that oxalic acid will impede the absorption of calcium. For example, spinach contains between 0.32 and 1.26 g per 100 g and is thus rightly known for its high content of this acid. More in-depth studies in tortoises would be highly desirable here. In human care, Egyptian Tortoises should be fed a diet low in protein, i.e., less than 20 %, particularly rich in raw fibre, and with an exceptionally high Ca:P ratio, which will be discussed in detail later in the present book.

All known populations of the Egyptian Tortoise in Libya are situated in a subtropical vegetation zone with a steppe climate that is moist and humid in winter and dry and humid in summer. These partly extreme conditions have enforced particular evolutionary adaptations in *Testudo kleinmanni* that have been allowing them to survive here. In fact, this species is adapted to living in one of the driest tortoise habitats of all.

An active Testudo kleinmanni amongst feeder plants in its natural habitat, Negev Desert, Israel

Photograph Aviad Bar

The altogether light colouration almost entirely without black markings of the dorsal shell facilitates a maximum reflection of solar radiation in an environment where shade will be scarce and everything very hot in summer. As a result, less radiation is absorbed and the body is less heated, reducing the risk of lethal overheating. The small body volume of the Egyptian Tortoise is another adaptation to its natural habitats, because the scant vegetation will provide little in terms of abundance of shelters, and what is available will be limited in terms of space. Without shelters, the animals would overheat very quickly in the hot environment, though (SCHLEICH 1989).

The size of the territory of *Testudo kleinmanni* differs significantly from those of other Mediterranean species of *Testudo*. The Egyptian Tortoise has a home range of ca. 16.4 ha (males) and ca. 7.2 ha (females) on average, i.e., males have territories about twice the size of those of females (GEFFEN & MENDELSSOHN 1988). This is in stark contrast to the average territory sizes of other Mediterranean tortoises, which range from 0.65 to 1.69 ha (males) and 0.74 to 2.41 ha (females), respectively, in which females generally have larger territories than males. A deviant maximum of 4.6 ha (males) and 7.4 ha (females), was found in a natural habitat of *Testudo hermanni hermanni* in which the availability of food was compromised (VETTER 2006). The latter also provides a plausible explanation for the much larger territory size of Egyptian Tortoises in that it is a result of a reduced availability of feeder plants. Like all species of *Testudo*, the Egyptian Tortoise exhibits a great site fidelity, and the individual tortoises will always return to their individual retreats.

The diel activity profile of the Egyptian Tortoise varies very substantially with the seasons, with a maximum travelling distance of 340 m having been recorded for males and one of 155 m for females. The tortoises are forced to wander long distances if they are to find enough food, which explains their reputation of being a particularly lively species. While they will be little active during the hot months of summer and leave their retreats only briefly, if at all, in the early morning and late afternoon hours, they are commonly active between 10 and 14 h in autumn, winter and spring (GEFFEN & MENDELSSOHN 1989, SCHLEICH 1989).

Its season-induced activity pattern is another adaptation to the particular climate in its natural habitats, rendering *Testudo kleinmanni* the only species of the genus *Testudo* to be most active in the winter months and to routinely observe a period of summer dormancy (aestivation).

In September/October, following the rainless months of summer, precipitation events slowly resume, plants start germinating or regrowing, and temperatures decrease to more moderate levels. This is when the Egyptian Tortoises will gradually resume their activity as well, having awoken from their interlude of aestivation. Their activity period extends from October through May, with the greatest level of activity being observed from December through March. Local weather conditions may allow them to become active earlier or continue to be so for longer, like until June/July, in places (Geffen & Mendelssohn 1989). With temperatures rising further from June and precipitation decreasing towards zero, both factors reaching their maxima usually in August, and their sources of food disappearing, they are forced into their summer retreats. The only way for Egyptian Tortoises to survive this period is by ceasing almost all activity and feeding, and retreating into their shelters for several weeks or months – they will typically aestivate from about June/July through September.

Thermoregulation during their activity phase, the need to escape the scorching summer temperatures, and the prevention of dehydration in either case make *Testudo kleinmanni* select retreats that will be optimal for each purpose separately (Attum et al. 2013, Buskirk 1990, Geffen & Mendelssohn 1988, Mendelssohn 1982, Schneider & Schneider 2008). The heat and dryness of summer sees them taking preferably to rodent burrows for aestivating where they will find a relatively low and constant temperature at raised and fairly constant levels of relative humidity and ambient moisture. In contrast, they will rather hide in shelters under bushes (*Artemisia monosperma*), in winter and spring, where they can half bury themselves in the sand. Here, Egyptian Tortoises prefer the east- or south-facing sides of a bush, which will enable them to heat up faster after sunrise (Geffen & Mendelssohn 1989). Depending on the weather conditions during their activity phase, they will select smaller bushes when temperatures are moderate, and larger ones of at least one meter tall when temperatures are hostile. Below small shrubs of maybe 20–30 cm in diameter, temperatures differ rather little from those all around, but larger shrubs from about 50–80 cm in diameter will provide relatively stable values of ca. 30 °C in summer, ca. 19 °C in spring, and ca. 17 °C in autumn, even when peak values of at maximum 54.5 °C and at minimum 3 °C (Attum et al. 2013). Rodent burrows and larger shrubs thus offer microhabitats with more stable and therefore safer conditions, facilitate the required thermoregulation, provide protection against dehydration, and thus constitute suitable refuges that ensure the long-term survival of these tortoises.

Egyptian Tortoise buried in the sand beneath bushes during their activity period in winter and spring.

Testudo kleinmanni in its natural habitat, Negev Desert, Israel

Photograph Aviad Bar

Unlike other species of tortoises, *Testudo kleinmanni* in its extreme environment cannot take in enough water by accessing puddles, rivulets or the like. Even heavy downpours will see the water disappear into the ground very quickly so that no puddles can form that would allow the tortoises to drink from. Egyptian Tortoises appear to therefore cover their need for water entirely from their feeder plants and by sucking up drops of rain and dew that will form on plants and on their shells (Schneider & Schneider 2008).

Comparatively very little is as yet known about sexual maturity, mating, oviposition, incubation parameters, and hatching in the Egyptian Tortoise in its natural environment. Females become sexually mature at an age of about five years, males probably earlier, at three, but usually at four years (Vetter 2002). Field studies on the reproductive biology of *Testudo kleinmanni* in the northwestern parts of the Negev Desert allowed Geffen & Mendelssohn (1991) to estimate the age of egg-laying females at between six and nine years, with age estimates being based on counts of growth rings around the scutes of the dorsal shell, and carapace lengths of reproductively active females ranged between 107.3 and 120.7 mm (Attum et al. 2011).

In March, when both temperatures and the hours of sunshine per day start increasing, the habitats will be blanketed with a sea of flowers. This time of abundant food supports the mating season of the Egyptian Tortoise (Lortet 1887, Geffen & Mendelssohn 1991). Courtship consists of the male ramming the shell of a female in an attempt to keep her in a spot and mount her. This effort is accompanied by the male emitting high-frequency sounds that are difficult to describe in words, being a mix of cooing, trilling and tweeting (Flower 1933, Loveridge & Williams 1957). Oviposition events can then be observed in May and June. To this end, the females will excavate shallow pits of some 3–4 cm deep in a shaded spot on the south-facing base of a bush. Once the eggs have been laid, the pit is closed with sand. Sometimes the female will wet the soil with some urine, which may for one mask the scent of the eggs against detection by predators and, for the other, increase the moisture level within the nest pit or even make digging in the substrate easier. Most clutches comprise two, some only one, and few as many as three eggs. Two or three clutches are deposited at intervals of 20–30 days each. Substrate temperatures measured in the wild at a depth of 5 cm were 24.3–38.2 °C in July, 23.5–35.3 °C in August, and 23.9–34.5 °C in September. Moisture levels of the sand at a depth of 5 cm were calculated by weighing samples that were then dried and weighed again. The difference was found to lie between 0.16 and 0.27 %. Given these conditions, incubation took about 70–90 days, but only a very few instances could actually be monitored in the wild thus far. Hatchlings eventually emerge from August to September (GEFFEN & MENDELSSOHN 1991), when they can benefit from the return of the rains and the associated increase in food.

Kept in human care, times of mating and oviposition, incubation periods, and the point of time of hatching have been noted to deviate in part substantially from corresponding data from the wild (see p. 188).

A Testudo kleinmanni female depositing her eggs

Photograph: Franziska Biedenweg

Captive Care

Fundamental Requirements

Our husbandry practices and strategies for the Egyptian Tortoise are guided as far as possible by realisations of its life in its natural habitats, as described in the previous chapters. The greatest challenge here is that these tortoises are adapted to living in an extreme semidesert climate that will force them spend the dry and hot time of summer in a state of dormancy known as aestivation. Considering our own Central European climate with its normally comparatively cool and rain-rich summers, it will therefore be practically impossible to keep *Testudo kleinmanni* permanently in an outdoor enclosure. There are, however, reports on housing these animals at least temporarily in outdoor pens, portable boxes, or in hotbeds set up on a terrace, or even letting them access an outdoor enclosure from their actual pen in a conservatory (Connor 1993, 1994, Dworschak 2012, Hobbs & May 1993, Hoekstra 1994, 1996, Reinarz 2008, Scheffel 2011 a, b). Most commonly, these tortoises are kept in terraria, though (Branser 2011a, b, Fritzsche & Fritzsche 2002, Gad 2000, 2001, 2005, Gruber 1993, 1998, Highfield & Martin 1995, 2001, Scheffel 2014, Schmidt 2009, Stettner 2004, 2006, Wittmund 2005), more rarely in large open-top pens in a conservatory, greenhouse or dedicated attic room (Dworschak 2010, 2012, 2013, Fritzsche & Fritzsche 2002 Reinarz 2008, Hebbeler pers. comm.).

What you should know: If you are planning on keeping Egyptian Tortoises you should be aware of the fact that the equipment required for creating optimal husbandry conditions will be rather expensive. It will not only be the costs for acquiring a terrarium and all the technology needed for operating it, however, but also substantial running costs for electricity in particular that exert strains on the monthly budget for many years to come.

This chapter will in the following provide information on necessary technological equipment and accessories, illustrate various alternatives, and describe concrete examples of husbandry strategies. The information and descriptions presented here do by no means claim to be the only "proper" approach, but merely offer possible options. As always, particular problems require particular solutions, and various factors will depend heavily on individual spatial and other conditions that every keeper will have to take into careful consideration for him- or herself. In the end, every keeper will therefore have to question his or her practices through experimental trials of alternatives, variations and possible changes to see whether behavioural responses by the tortoises indicate that their lives would be improved by a particular change in the setup.

The Terrarium

Materials

Several options lend themselves when it comes to selecting a suitable terrarium. It starts with the material from which the terrarium is made: There are models constructed entirely from glass, those that are fashioned from wood and have glass panes, and even those that are cast or moulded from plastic. You also have a choice between fully closed models, those with an open top, or others with a partially open top. All these alternatives come with their individual advantages and disadvantages. Your choice will lastly very much depend on your personal situation, the place where the tank is to be set up, and the abiotic factors at this particular spot during the course of the year. All these variables inevitably have an impact on the temperatures and humidity values that will have to be reliably controlled later on. For example, it may prove more difficult in a terrarium set up in a living room to achieve the necessary low night temperatures in winter than in a tank that has been placed in an unheated room. On the other hand, it may prove difficult to achieve the required high summer temperatures in an open-top terrarium in a living room. Selecting a suitable spot for placing the terrarium must therefore be part and parcel of your planning.

Terraria can today be bought ready-to-use in a wide selection and from a multitude of suppliers both on the Internet and in specialist shops. You may alternatively want to go for an assembly kit, or, if your DIY skills are reasonably developed, plan and custom-make your tank according to your personal requirements.

What you should know: No matter which type of terrarium you may select, you will need to ensure that ventilation is not limited to the top, but also is effective at the ground level at which the tortoises actually live. This means that there must be ventilation surfaces near the ground as well. Draughts, on the other hand, which are a particular risk in the case of open-top tanks, must be avoided by all means possible.

Glass terraria: Daylight falling into a terrarium with glass walls may help to increase the basic level of brightness in the tank if it is positioned near a window. Here, it must be taken into consideration that placing it at a south-facing window may create a risk of overheating from insolation that may be just too intense. This risk is obviously greatest in closed glass terraria, especially when heat lamps are operated in addition. The glass panes may furthermore have the effect of a burning-glass. The advantages of a glass terrarium clearly are that it can be cleaned easily and thoroughly and even disinfected if and when necessary. A distinct disadvantage is that glass is a good thermal conductor, which will cause lots of heat to be lost to the surroundings that has to be compensated for by more electrical energy for heating the tank to the desired temperatures and keeping them there. The required heat lamps cannot be mounted simply by screwing suitable brackets on, but will require special mountings. Finally, tortoises have no concept of glass, will not see it as an obstacle and continue to try to break through it if it has not been covered at their level and thus made opaque. They will also be aware of all movements in the room around them and may be confused by their reflections on the glass surfaces.

Wooden terraria: Terraria made from wood are much lighter in weight compared to those made from glass and provide better insulation, which will allow you to save on your electricity bill. Mountings, brackets and fittings can be easily affixed and just as easily relocated if required. Building a wooden terrarium is a fairly straightforward affair even if your DIY skills are not really at master level, costs will be moderate, and fitting it into a given space is much easier than when you have to work with glass. If you need building plans, these can be found in their numbers on the Internet, and the required front panes of glass and their railings are available from the terrarium supply trade. Orientated Strand Board (OSB) is most commonly used for this purpose, more rarely texture coated board or glued laminated timber. Ready-to-use wooden terraria or self-assembly kits from commercial sources usually also make use of OSBs. Depending on the positioning of the tank, panes of glass may be limited to the front, or you may want to replace the one or other wall with glass as well, e.g., if the terrarium is to serve as a partition in a room and is positioned freely. The rather drab OSB rear and side walls may be pasted with a landscape poster or structured panels of the type available from commercial suppliers, depending only on your personal taste and extent of imagination. Doing so will have no bearing whatsoever on the quality of life of the Egyptian Tortoises living in the terrarium, but only satisfy your sense of aesthetics.

Top: A wooden terrarium under construction
Bottom: With a photo rear wall
Photographs Ricarda Schramm

Top: A terrarium made from OSB *Photograph Joe Stammet*

Bottom: Sealing with pond liner *Photograph Ricarda Schramm*

Before a wooden terrarium can be filled with bottom substrate, its bottom section will first have to be sealed against moisture. This can be done by painting it with an epoxy sealant, coating it with liquid pond liner, or by covering it with sheet pond liner (preferably made from EPDM, ethylene propylene diene terpolymer). If you opt for a paint-on method, you may want to additionally seal all seams with aquarium silicone afterwards, which will of course not be necessary in the case of sheet pond liner. The latter needs to be folded neatly into the corners, pulled up to the height of the substrate that is to follow later, and simply stapled to the wooden walls.

A disadvantage of wood is that it is less easily cleaned.

Plastic terraria: Plastic terraria are either made from so-called Forex® sheets, which are very light-weight, white, mould-foam, PVC (polyvinyl chloride) sheets, or from hard polyurethane foam (PUR). These can be worked with very much like wood. The advantages of these plastics include their reduced weight, ease of working with, ease of keeping clean and disinfecting, their resistance to moisture, and their being a better insulator compared to both glass and wood. Fittings for lamps etc. can be simply screwed on here as well. The mentioned materials are furthermore fire-retardant, which makes operations safe to some extent. Forex® sheets are relatively pliable and therefore fit only for smaller terraria, whereas those made from PUR are much more rigid. The sheets themselves do not have to be protected from moisture, but you should take the pains of sealing all seams with silicone to prevent moisture from leaking out.

According to manufacturer's specifications, the two plastics mentioned are UV-resistant, but as to how far this is a valid claim in the long run remains to be seen.

Dimensions

Legal minimum requirements or recommendations for the minimum dimensions of terraria exist in Germany, Austria and Switzerland, but may in future also come into effect in other countries. These are based upon expert opinions and meant to ensure that animals are kept in spatial conditions that will enable them to live according to their biological needs. *Testudo kleinmanni* did (unfortunately) not feature in the edition of the German legal minimum requirements for reptiles of 10. Jan. 1997 that was valid when this text was penned. Guidelines for the minimum dimensions of their terraria were given only by inference from those for all other species of *Testudo*, even though Mediterranean tortoises are actually best kept in outdoor installations and not in terraria in the first place.

Be that as it may, using the values suggested for *Testudo graeca* for *Testudo kleinmanni* instead, the following minimum dimensions are calculated based on the carapace length of the largest specimen times eight and its carapace width times four. The result is then applicable to the joint housing of two adult specimens, and 10 %, must be added for a third and a fourth one, and 20 % for a fifth. This means that a terrarium for keeping two adult Egyptian Tortoises with carapace lengths of 12 cm must at least measure 96 x 48 cm in floor space, which would not even amount to a full 0.5 m². Such tank would need to be only 142 x 71 cm or ca. 1 m², even for five adult *Testudo kleinmanni* which is decidedly way too small for these very active tortoises. German authorities are at present busy revising their guidelines.

While *Testudo kleinmanni* is named in the Swiss Ordinance for Animal Welfare of 23. Apr. 2008, the same formula as in Germany must be applied.

A different legal situation is encountered in Austria's Law Gazette II, No. 486 of 17. Dec. 2004 in which *Testudo kleinmanni* is not only listed, but according to which other minimum terrarium dimensions are to be provided:

Size / Number	**up to 6 cm in carapace length**	**6-12 cm in carapace length**	**more than 12 cm in carapace length**
1 - 2 specs.	0.50 m²	1.00 m²	2.00 m²
For each additional spec.	+0.20 m²	+0.40 m²	+0.50 m²

This means that a colony of five adult tortoises with carapace lengths of 12 cm should have available a floor space of at least 2.2 m² which finally comes close to what Egyptian Tortoises need in terrarium space.

What you should know: A terrarium for the active Egyptian Tortoise can never be too large – only too small.

With larger terraria responding much slower to climatic fluctuations, the right climate will be more easily established and maintained.

Even if you first acquire small juveniles, these will prove to be particularly active and relatively fast-growing. It is therefore recommendable to size their terrarium right from the start according to the number of adult Egyptian Tortoises it is supposed to accommodate in the end.

An impression of the active lifestyle of Egyptian Tortoises and their home ranges in their natural habitat is provided on p. 60 (Geffen 1987, Geffen & Mendelssohn 1988, 1989, Gruber 1998).

Bottom Substrates

In the wild, *Testudo kleinmanni* llives mainly on compacted sand within a semidesert landscape or in depressions between sand dunes. This aspect should therefore provide the cue for the selection of the best bottom substrate for a terrarium for these tortoises. Generally speaking, the bottom substrate should be firm enough to walk on and dig into, water-permeable, moisture-retaining, and not tend to produce excessive dust. In nature, rains do not normally cause puddles to form, but rather will the water seep away immediately so that the surface of the ground is dry once more soon afterwards (Schneider & Schneider 2006, 2008). Life-sustaining dampness and humidity remain accessible in the underground shelters of the Egyptian Tortoises, however, for which purpose they will utilise rodent burrows or bury themselves in the sand beneath bushes or roots. Here, they will encounter the moister microclimate that actually enables them to survive in the given semidesert climate in the first place, in particular in summer. The bottom substrate should be compact enough when dry to permit walking on it without sinking in, but loose enough to also permit digging when damp as to enable gravid females to excavate nest pits whose walls will not collapse all the time. These tortoises spend most of their life in close contact with the bottom substrate. Selecting one for their terrarium that is less than optimal may pave the way to health issues such as fungal infections if it is too moist, the growth of humped shells if it is too dry, or respiratory problems if it is too dusty. A. C. Highfield (2001–2008) provided an overview of the advantages and disadvantages of various bottom substrates and arrived at the conclusion that coir (coconut humus) would be unsuitable because it may produce dust, tends to dry out completely, may be ingested with food and then cause intestinal problems, does not offer any thermoregulatory advantages, and actually creates a perfect environment for the proliferation of moulds, parasites and pathogens. Highfield also advised against the use of cedar or pine bark or substrates that were based on these, as these would release toxic vapours, in particular at high temperatures, could cause intestinal obstructions if they were ingested with food, and also create an ideal substrate for parasites and germs. Added to all these is the fact that all these substrates have an acidic pH. In the case of *Testudo kleinmanni*, he recommended a sand/loam mix made from washed playground sand and powdered loam at a ratio of 70 % sand and 30 % loam, which corresponds to a proportion of 2.3:1, and means a rather high content of loam. In a previous publication, HIGHFIELD & MARTIN (1995) had even given a mixing ratio of 60 % loam and 40 % sand. A 1:1 mix of sand and loam is used by Hans-Ulrich SCHMIDT (2009), whereas GAD (2000, 2001, 2005) utilised pure sand that was mixed with loam only at the intended oviposition spot in order to provide the require stability for excavating proper nest pits.

Most keepers of Egyptian Tortoises use a sand/loam mix as a bottom substrate, mixing it from washed playground sand and powdered loam from the terrarium supply trade, commonly at a ratio of about 4:1 or even 5:1 (DWORSCHAK, 2010, 2012, REINARZ 2008, SCHEFFEL 2014, WITTMUND 2005). Ready-to-use mixes are by now also available from specialist suppliers.

The mixing ratio of playground sand and powdered loam

Photograph Franziska Biedenweg

A bottom substrate mixed from playground sand and loam in a terrarium

Photograph Franziska Biedenweg

Another option, which rarely features in forums or on webpages dealing with the captive care of reptiles, should be contemplated when selecting a bottom substrate for the keeping of *Testudo kleinmanni*. This is because one must ask oneself why particles including loamy ones should first be washed out of the sand, only to add loam powder afterwards? A reasonable alternative here exists in the shape of building sand of the type used for bricklaying, i.e., sand in the condition it will have come out of a sand pit. Sand for construction works usually already contains up to 3 % loam, which will lend it the desired attributes at least to a certain extent. This loam content will still not be large enough, though, so that building sand also needs to be fortified with some more loam powder. In order to achieve a mixing ratio of 5:1, which would require adding 200 g loam powder per 1 kg playground sand, the 30 g contained in 1 kg building sand need to be considered so that only 170 g of loam powder have to be added. Calculated in litres, and given a specific weight of 1.5 t/m^3 for building sand, 10 l (1 bucketful) of building sand weigh 15 kg and 2550 g loam powder have to be mixed in. It is entirely unnecessary to work to the exact gram, however.

Building sand is available from building supply outlets and will typically have a

Mixing ratio of building sand/loam powder for 10 l of building sand

Photograph Ricarda Schramm

reddish colouration. Loam powder can be found in specialist shops and comes in a range of colours.

The depth of the bottom substrate in the terrarium should be some 8–10 cm in order to provide a moisture gradient into which the tortoises can dig for sheltering, enables females to excavate proper nest pits, and allows you to sink in plants.

The amount of sand needed is calculated from the surface area of the terrarium (i.e., length x width) multiplied with the desired depth of the substrate.

Once the terrarium has been placed in its intended spot, the sand is poured into it first and thoroughly moistened through with water. Then the loam powder is added, and everything is mixed together until a homogeneous mix is obtained.

This mass can then be sculpted, enabling you to create a relief that will remain relatively stable after it has dried.

What you should know: It is important to mix sand and loam not when dry, but rather by adding water in order for the two components to become thoroughly tied. Once the bottom substrate has been distributed and the terrarium set up, the ground needs to dry for several days or even weeks before any Egyptian Tortoises can move in.

The substrate now conforms to all desired requirements: No water will remain standing following a raining session but dissipate quickly, the dried surface will be firm enough for the tortoises to not sink in, it will not produce dust, it will allow digging when damp, moisture will be retained in the deeper layers and release it slowly, and females will be able to excavate stable nest pits. The bottom substrate will thus very closely emulate the conditions in the natural habitat.

Bottom substrate mixed from building sand and loam in a terrarium

Photograph Ricarda Schramm

The bottom surface is kept clean by collecting faeces and leftover food every day. Scattered particles can be effectively collected by means of a hand-held vacuum cleaner. Replacing the bottom substrate completely will thus become necessary only after a few years or when disinfection becomes a necessity following a health problem amongst the tortoises. If you deem it necessary you can scrape off and remove some of the top layer and refill what is missing with fresh substrate. Because every little change in the environment means stress for *Testudo kleinmanni*, replacing their substrate should be limited to when it will really be necessary and then preferably step by step.

A juvenile Testudo kleinmanni in a terrarium with a bottom substrate mixed from building sand and loam

Photograph Ricarda Schramm

Lighting

General Considerations

Lighting terraria and which lamps are best for what are topics terrarium keepers have been discussing and arguing about with passion for ages and this subject has filled entire books and subject journals. An aggravating fact is that the supply industry offers uncounted options for illuminating terraria, which the non-expert will find difficult to compare and use for making an informed choice. The framework of the present book does not permit to discuss this complex subject in all its facets comprehensively either or even list all available options, especially considering that technology in this field develops at great speed and new products enter the market all the time. We have therefore limited ourselves to providing merely an introduction to some fundamental technical details and describing a few suitable lamps with the aim of bringing at least some "light into the dark" and by that make the selection process easier. As yet, there is no such thing as the "perfect lamp". Defying all technological progress, all lamps presently available come with advantages and disadvantages so that there will be no way around installing different sources of light in the terrarium, because just one type will not be sufficient.

Housed in a terrarium, Egyptian Tortoises cannot enjoy the benefits of natural sunlight. It is therefore crucial that lighting conditions for them be optimised by selecting a combination of suitable lamps. Here, brightness, colour temperature, colour rendering, and the spectrum covered by the lamps all play important roles. Comparisons must always be based on measurements of natural sunlight taken in the wild and the goal must be to come as close to these as possible, which is also the aim of the research undertaken by the manufacturers of these lamps.

Note: The following paragraphs discuss the fundamentals of light and some technological details of lighting. If you find these discussions too technical and/or detailed, you will find a brief summary on page 104.
If you, on the other hand, are interested in more information, you are referred to the special-subject literature quoted in key places.

Brightness: The distribution range of *Testudo kleinmanni* lies in regions covered by semidesert. There, the sunlight is much more intense than at our latitudes, and the number of hours of sunshine will be much greater, amounting to some 3,000 hours per annum in Libya as compared to about 1,300 to 1,900 hours per annum in Germany. It is therefore of utmost importance to provide terrarium-kept tortoises from these regions with light of corresponding brightness for a corresponding number of hours per day.

The brightness of light is measured in Lux (lx), even though other units of measurement may also be used regionally. The unit of measurement, Lux, indicates the volume of light (in Lumen, lm) that hits an area (m^2). In other words, the volume of light (lm), i.e., the amount of light that is emitted by a source of light, is related to an illuminated area, with 1 lx equalling 1 lm/m^2. About 100,000 Lux are measured on a sunny day. A similar degree of brightness is reached only in modern operation theatres and impossible to attain even remotely by means of the lighting systems available for terraria. With the distance between Sun and Earth being as large as it is, the possible change in the Lux value through shortening the distance by what is possible on Earth will be infinitesimal. This situation is entirely different when it comes to the lamps above the bottom substrate of a terrarium. Here, the distance is comparatively very short, and changing this distance will result in a major change of the brightness value. This means that the value of brightness is variable in a lamp, because it depends on the flow of light (Lumen), on the reflector of the lamp, and the distance of the lamp to the illuminated area. Direct comparisons of various lamps on the basis of the data provided on their brightness will therefore be hardly possible. Values on the flow of light (lm) or the yield of light in Lumen per Watt (lm/W) would render comparisons more informative, but, alas, these are rarely found in the product information provided by the manufacturers. Here, one should always keep in mind that "A lamp has no Lux – it will produce Lux" (Geissel & Kober 2015).

When selecting lamps for the basic illumination you should as a matter of principle preferentially opt for those with the highest possible values of brightness (lx), the highest light flux (lm), or the best yield of light (lm/W).

Colour temperature: The colour temperature specifies in a figure the impression of the colour of the light emitted by a lamp. It is measured in Kelvin (K). The colour temperature of the light in a terrarium should likewise come as close as possible to that of natural sunlight in the wild. Here, you should be aware that the spectral ranges of lamps may vary significantly even if they emit light with the same colour temperature.

Based upon how the human eye perceives colours, certain sections of the colour temperature are grouped under more generally informative terms: "warm white" at less than 3,300 K, "neutral white" at 3,300–5,000 K, and "cool white" or "daylight" at more than 5,000 K. For comparison, incandescent lamps emit light in the range from 2,600 to 3,000 K, "neutral white" fluorescent tubes around 4,000 K, moonlight around 4,120 K, and sunlight from a clear sky around noon is measured at between 5,500 to 5,777 K. Lamps of the "daylight" type are those emitting light of between 5,300 and 6,500 K out of a specially adapted spectrum.

Colour rendering: Colour rendering is optimal under the full spectrum of sunlight. To specify how a lamp renders colours, a colour rendering index, R_a, is provided, which reaches its maximum of 100 in uncompromised sunlight. This clearly suggests that the colour rendering index should be as high as possible when selecting lamps for the terrarium. Here, you need to keep in mind that lamps with the same colour rendering index may have different spectra, which is an important factor in the selection process as well.

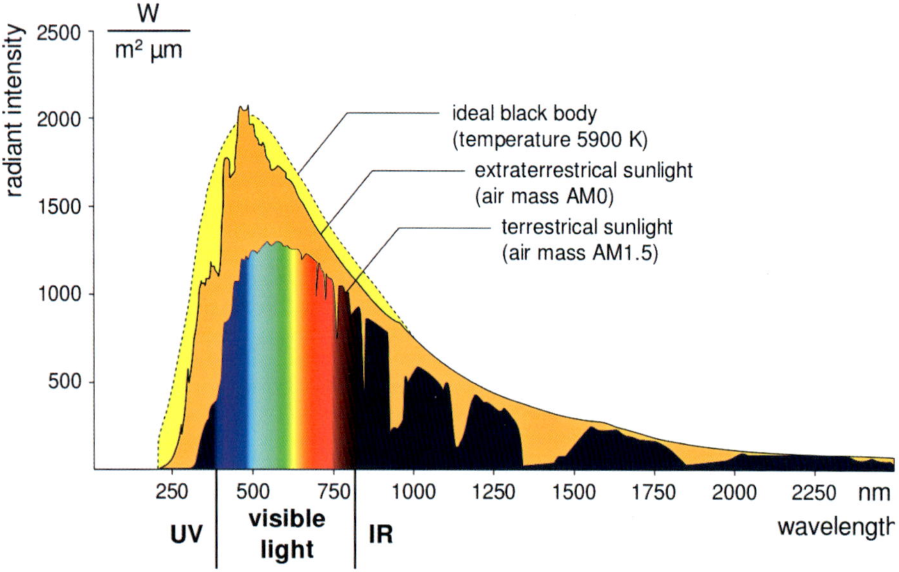

Spectra and intensities of solar radiation at AM0 (space near Earth) and AM1.5 (around the apex of the sun's motion in the German city of Karlsruhe) compared to the emission of a perfect black body at a temperature of 5,900 K. The black body represents an idealised source of thermal radiation and is used, e.g., as a theoretical model and reference for practical studies of electromagnetic radiation.

Spectrum: Light forms part of electromagnetic radiation, and optical radiation incorporates UV radiation, visible light, and infrared radiation. The individual ranges are measured by the wavelength of this radiation and measured in nanometres (nm). Light that is visible to the human eye ranges between about 380 and 750 nm. The individual ranges of wavelengths produce certain colours, and adding all of them together will produce the impression of white. The colours of objects, on the other hand, are created by their absorbing parts of the spectrum and reflecting others. The impression of a colour is the result of the intensity at which certain wavelengths are represented.

The human eye contains on its retina, next to rods for seeing in dimmed-light situations, three types of cones that are agitated by incoming light in the ranges red, green and blue and thus enable us to perceive colours (trichromatism) (Jacobs 1996). In contrast to the human eye, the retina of reptiles possesses four types of cones that are sensible to red, green, blue and ultraviolet (tetrachromatism), which enables them to additionally perceive spectral light at about 320–380 nm, i.e., in a range known as the UV-A part of the spectrum. This ability has been demonstrated in various species of chelonians both by means of direct examination of the retina (Ammermüller & Kolb 1996, Loew & Govardovskii 2001) and behavioural-response experiments (Arnold 1986, Arnold & Neumeyer 1987, Kelber et al. 2003, Pelitteri-Rosa et al. 2010, Simang et al. 2010).

Colour vision in the UV-A range can play a role for reptiles in finding a mating partner, i.e., when certain cues such as colour signals are broadcast during their mating season. Some reptile species are furthermore known to convey certain states of disposition by changing their body colours (e.g., chameleons). Seeing in this range may also help with finding food, because flowers and even some fruits will reflect UV light and so stand out through their colours in the UV-A range of the spectrum. This ability is not unique to reptiles, though, but has also evolved in birds and bees that likewise possess UV receptors on their retinae (Hempel et al. 2014, Kelber et al. 2003) and are in this manner able to detect flowers that depend on attracting them for pollination. These flowers are perceived by them in colours that will be entirely different to what they appear to the human eye (Chittka et al. 1994).

The peaks of absorption of the four types of cones differ to some minor extent in their ranges of wavelengths between various animal species with this ability.

Mixed-colour experiments revealed differences in colour vision between tortoises and humans. The human eye consolidates individual colours to form a colour hue, and the same impression of a colour can be obtained by mixing

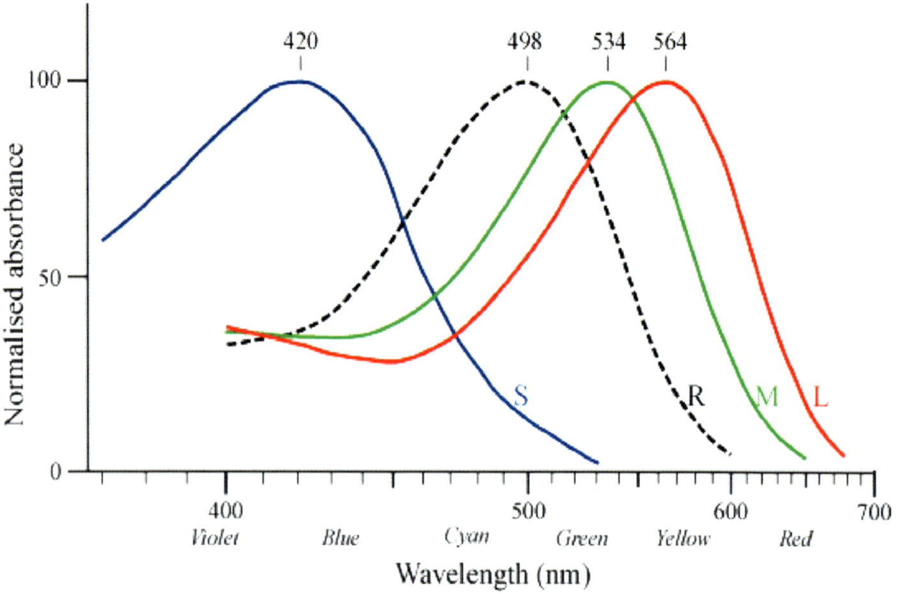

Top: Pigment absorption in humans (trichromatae); rods (dappled black) and three types of cones (blue, green, red)

Bottom: Pigment absorption in tetrachromatae (bird); four types of cones (UV-A, blue, green, red)

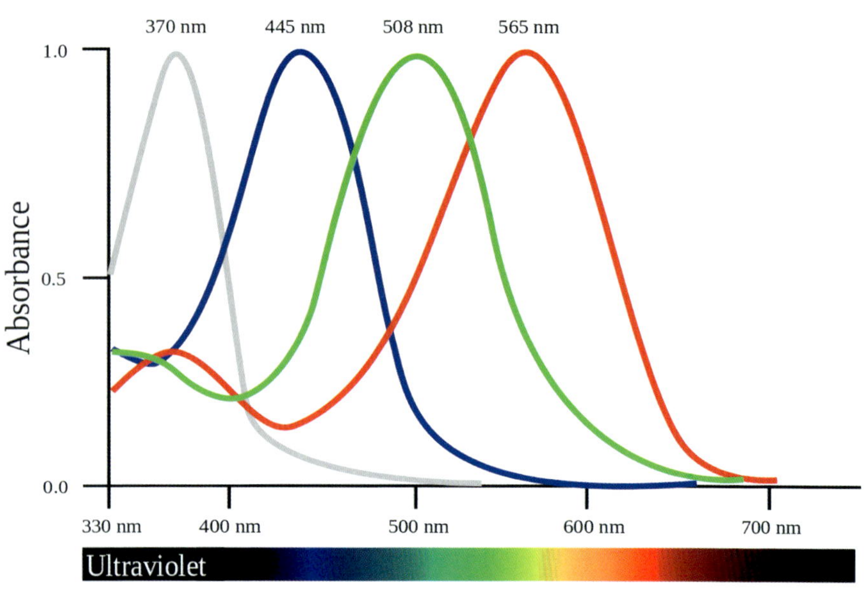

various wavelengths together. This is not exactly the same in animals with the ability to see in the UV range. When a behavioural experiment trained Red-eared Slider (*Pseudemys scripta elegans*) to a wavelength of 404 nm (violet) it showed that – in contrast to humans – it was impossible to create the same impression of this colour by mixing 450 nm (blue) with 683 nm (red). On the other hand, these turtles were unable to distinguish a mix of 450 nm (blue) and 367 nm (UV-A) from 404 nm (Arnold 1986, Arnold & Neumeyer 1987).

Basic Illumination

The basic illumination in a terrarium serves to define the photoperiod, i.e., the number of hours of daylight per day. It does not take care of providing the tortoises with warmth or vital UV-B radiation. The photoperiod needs to be adjusted according to climate charts every month and then left to be managed by timers. If not all lamps are switched on/off at the same time, approximations of dawn and dusk can be created. An even better solution may lend itself in the shape of a dimming timer that is left to control dimmable lamps to this effect.

Terraria are typically illuminated with fluorescent tubes, halogen metal vapour lamps (HQI), and, since recently, also with LED lamps. The types of lamps listed here can all be combined for creating basic illumination in terraria.

Fluorescent Tubes: Fluorescent tubes have been used for lighting terraria for decades and are still highly suited for creating the basic illumination. These are low-pressure gas-discharge or metal vapour lamps that are filled with mercury vapour and lined on the inside with a fluorescent coating.

Their different colour emissions are the result of mixing various luminescent substances.

Best suited for use in a terrarium housing *Testudo kleinmanni* is light with a colour in the range of 5,500 through 6,500 K. Animated by an electrical current, the mercury vapour will produce UV radiation inside the tube that is for the largest part converted into visible light by the fluorescent coating and to a small extent left as UV-A- and UV-B radiation. The UV-B radiation will not be able to penetrate the glass of the tube unless it is of a type that is specifically permeable for it.

Most widely used today are fluo tubes of the T8 or T5 types (T = Tube), They differ in diameter, with the older T8 measuring 26 mm und the slim, newer T5 only 16 mm. The fittings of these tubes are standardised, named G13 for T8 and G5 for T5, so that these names together with the length of the tube will allow you to select right-size fittings for your tubes. Fluorescent tubes require ballast units ("transformers") that will take care of igniting the plasma within the tube at a high gate trigger voltage and then limit the current strength (amperage). The conventional ballast units that used to be notorious for their producing flickering at 100 Hz have since been replaced with electronic control gear (ECG) that works at much higher frequencies and thus is flicker-free, has an improved light yield, and uses up less electricity for itself. It therefore makes great sense to only use fluo tubes that are run via ECG. The maximum light flux of T5 tubes is reached at a temperature of 35 °C and no longer at 25 °C (T8), for which reason they are more commonly run at near full capacity in a terrarium and produce a higher yield (ARTH 2012, GEISSEL & KOBER 2015, SAUER 2004, WUNDERLICH 2017). Two types of T5 tubes are distinguished: T5 HO ("High Output") and T5 HE ("High Efficiency"), with a HO-type tube consuming more electricity, but also giving off more light than an HE-tube of the same length (ARTH 2012).

Fluorescent tubes do not have a continuous spectrum, but emit their light in bands (peaks) that correspond to the emissions of the luminescent substances contained. Three-band fluorescent tubes thus have one peak each in the wavelengths red, green and blue that will add up to white light. An advanced novelty is five-banded fluorescent tubes in which a minimum of four luminescent substances render the emitted light more similar to daylight and more continuous with fewer gaps in the spectrum. They are therefore also referred to as full-spectrum fluorescent tubes. Three-banded tubes have a colour rendering index, R_a, of 80 to 89, whereas five-banded types will reach 90 to 99. This is reflected in the standardised three-digit code used by the manufacturers to specify the colour rendering and colour quality of their products: The first digit indicates the range of the colour rendering index (8 for 80 to 89, 9 for 90 to 99), while the last two digits give the colour temperature without the last two zeros. For example, the specification "965" on a fluo tube means that it is of the five-banded type with an $R_a > 90$ (colour rendering index) and 6,500 K (colour temperature). The improved spectrum and colour rendering of five-banded fluorescent tubes is to the detriment of the light yield, Lumen/Watt, however (ARTH 2012).

A high colour rendering index of $R_a = 90$ or more alone is no clear indication of the tube also emitting radiation in the UV-A- and UV-B ranges (WUNDERLICH 2012b). This piece of information usually will have to be extracted from the manufacturer's product data sheet, which typically details the emitted spectrum including/or the volume of UV emissions in percent.

Some manufacturers produce full-spectrum fluorescent tubes with UV emission specifically for use in terraria, with examples being the UV-fluorescent tubes from Lucky Reptile (UV Sun T5 HO), Zoo Med (ReptiSun® 2.0, 5.0, 10.0 T8 and T5 HO), Exo Terra® (Repti Glo 2.0, 5.0, 10.0 T8), and others more. These tubes differ as to their emissions of UV light (all data adopted from manufacturers' specifications):

Manufacturer / Model	**Colour rendering index (R_a, CRI)**	**Colour temperature**	**UV-B/UV-A**
Lucky Reptile UV Sun T5 HO	90	6,500 K	No data
Zoo Med ReptiSun® T8, T5 HO ReptiSun® 2.0 (NatureSun 2.0 T8) ReptiSun® 5.0 ReptiSun® 10.0	98	6,500 K	1–2 % / 10 % 5 % / UV-A ? 10 % / 30 %
Exo Terra® Repti Glo T8 Repti Glo 2.0 Repti Glo 5.0 Repti Glo 10.0	98	6,700 K	2 % / 25 % 5 % / 30 % 10 % / 33 %

Corresponding UV-fluorescent tubes are also available from Arcadia, and all can be acquired with fitting operation gear including the necessary ballast units. Fluorescent tubes have a maximum light yield of about 80 lm/W (Lumen per Watt), even those that also give off UV light.

In order to illuminate the terrarium properly throughout, the lengths of the tubes and their arrangement should cover at least most of its entire length. Oliver Fischer (2016)) recommended installing one T5 HO tube per 30 cm of terrarium depth if the illumination consists exclusively of fluorescent tubes.

Sarina Wunderlich (2017) considered the UV content of UV-fluorescent tubes too large relative to the visible spectrum and UV-B too large relative to UV-A, because natural sunlight would have about 0.1 to 1 % UV-A, stating (quote, in transl.), "Or phrasing it in reverse: compared to their UV emission, fluorescent tubes are way too dark.". For his part, Markus GRIMM (2016) specified the distribution of wavelengths in natural sunlight as 6 % UV, 46 % visible light, and 48 % radiant heat, and his measurements taken with a spectrometer of the light emitted by fluo tubes often revealed excessive UV-A contents and unfavourable UV-A/UV-B ratios. He also noted, however, that both the brightness and UV radiation of T8 tubes would decrease very significantly from a distance of 10 cm to the lamp and therefore recommended using them only for low terraria.

The advantage of UV-fluorescent tubes clearly is that their spectrum will extend into the UV-A range and so enables reptiles to see colours in this range. They do not, however, facilitate a sufficient supply with UV-B, because their output of the respective wavelengths will be too small and be reduced further by the distance between the lamp and the tortoises in a large terrarium (DIEGEL 2010, FISCHER 2017/2018, WUNDERLICH 2012a). This problem is discussed in greater detail in the chapter on UV-B below.

Imperceptible to the human eye, the output of UV radiation will continually decrease the longer a fluorescent tube is operated, necessitating that it be replaced with a new one after about one year (GEISSEL & KOBER 2015, WUNDERLICH 2017, also manufacturers' recommendations), which will of course be a factor in the overall operating costs.

An inherent disadvantage of fluorescent tubes is the loss of yield due to their emitting light at an emission angle of 360 °. Some 50 % of the light are radiated upwards and therefore lost to the brightness at the bottom of the terrarium. This loss can be curbed by fitting the tubes with reflectors even though some of the reflected light will be directed right back at the tube and thus be lost by its casting a shadow. The better the reflector is fitted relative to the tube in both distance and size, the less light will be lost. Clip-on reflectors are typically angled in cross-section, which will disperse the light more than a curved model, but they are also typically at a short distance to the tube and will therefore reflect most of the light right back onto the tube. Parabolic reflectors in which the tube is mounted right in the focal point are more effective, because the larger distance from the tube will minimise the amount of light lost in the tube's own shadow; the light is dispersed to a lesser extent and reflected straight down (GEISSEL & KOBER 2015). A particular model of clip-on reflector promises an improved yield of light; it is doubly ellipsoid in shape, practically like two joined parabolic reflectors that will in this manner form a bridge along the exact midline of the tube (Trocal Longlife Power Reflect, Dennerle GmbH, Germany). This results in hardly any light being reflected back onto the tube, but the largest part of the radiation being redirected down instead. As a result of their slimmer shape, T5 tubes are less wasteful through shading than those of the more voluminous T8 type.

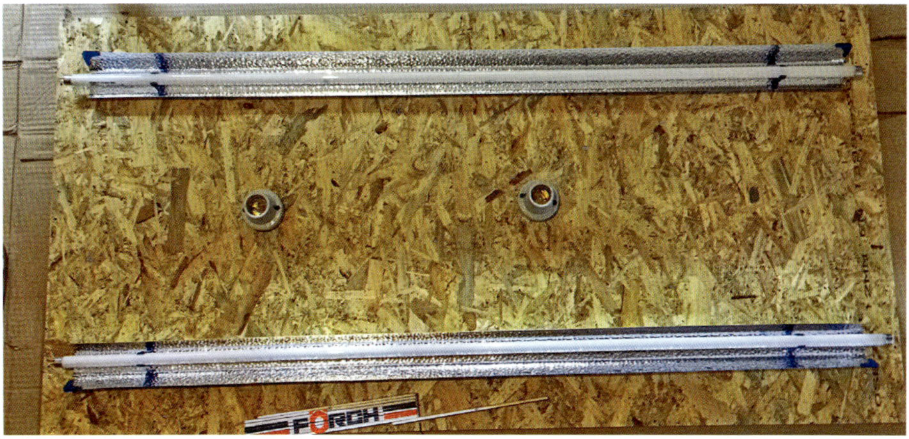

Mounting fluorescent tubes with clip-on reflectors to the lid of a terrarium; fittings for additional UV lamps can be seen in between.

Photograph Franziska Biedenweg

Overview: Fluorescent tubes should span across the entire length of the terrarium. If fluorescent tubes are to provide the full basic illumination, the width of the tank will determine how many of them need to be installed in order to ensure the desired brightness at the bottom. Using reflectors is obligatory to minimise the loss of light due to its dispersal upwards instead of down. An optimal setup can produce a yield of some 80 lm/W.
UV-fluorescent tubes enable tortoises to see colours in the UV-A range of the spectrum, but will not adequately cover their need for UV-B radiation. The latter must therefore be supplied by means of other lamps in addition.
The output of UV radiation will diminish as fluorescent tubes age, necessitating their replacement after about a year.

Halogen Metal Vapour Lamps: Halogen metal vapour lamps are high-pressure glow-discharge lamps, representing a technological advancement from mercury vapour lamps. They emit light that will come very close to natural sunlight with excellent colour rendering, high degrees of brightness and light yield, and they give off some of their energy in the shape of heat. The ratio of light to heat resembles that of sunlight. Light is in their case produced by means of halogenides (salts). By changing the mixing ratio of these salts, light of different colours and different colour rendering attributes can be produced. The supply trade offers halogen metal vapour lamps like those named CDM by Philips, and HQI (quartz rods) or HCI (ceramic rods) by Osram, for example.

Next to LED lamps, halogen metal vapour lamps are amongst the most effective lamps. They reach colour rendering indices (R_a) of up to 90 and light yields of up to 117 Lumen per Watt (lm/W) at good, and up to 110 lm/W at very good colour rendering. The illumination provided by them is more spot-like in character and less widely dispersed for which reason several of them are typically used for lighting a terrarium.

A built-in halogen metal vapour lamp (HQI) in a terrarium, on the left in the photograph.

Photograph Ricarda Schramm

The relatively high output of heat of halogen metal vapour lamps may cause ambient temperatures to be too high for *Testudo kleimanni* in closed-top terraria during the winter months.

A distinction needs to be made between the older quartz- and more modern ceramic-rod type burners in that ceramic-rod burners will maintain the colour of the emitted light throughout their life whereas it tends to shift towards green in the older quartz-rod burners with increasing age. Ceramic-rod burners also have a greater light yield of up to more than 100 lm/W (Lumen per Watt), whereas quartz burners may reach only a maximum of about 80 lm/W. The downside of ceramic burners may be that they will produce colour temperatures of only up to ca. 4,200 K whereas the quartz-rod types can reach values in excess of that (Fischer 2017/2018, Geissel 2016, Geissel & Kober 2015, Wunderlich 2012b).

The cover glass of classical halogen metal vapour lamps is permeable to UV-A radiation, but not to UV-B. This situation is different in so-called UV halogen metal vapour lamps (UV-MH), which are discussed in detail on page 93.

Overview: Halogen metal vapour lamps produce light that will be very similar to natural sunlight, have excellent colour rendering attributes (R_a up to 90) and a high light yield (up to 117 lm/W).
Lighting a terrarium equally brightly throughout will usually require the installation of several lamps. The relatively high emission of heat may cause overheating in closed-top terraria.

LED: This widely used term is short for "light-emitting diode". White light is usually produced by a blue diode whose spectral emission is converted into longer wavelengths by means of luminescent substances, quite similar to fluorescent tubes.

LED lamps have been employed in the aquarium hobby for quite some time, but their use as the source of the basic lighting in terraria is still a widely and hotly debated subject in both subject literature and Internet forums. While some think of them as entirely unsuited (Wunderlich 2012b, 2015, 2017), others have been reporting on positive experiences with LED lamps in terraria, and their numbers have been increasing (Akeret 2013, 2015, 2017, Geissel 2015, Geissel & Kober 2015, Grimm 2016, Kunz 2015, Schramm 2016). With LEDs having been used in terrarium applications only for a short time, long-term observations are still unavailable. Kriton Kunz (2015) portrayed a number of positive reports by renowned terrarium experts who have made practical experiences with the use of LED lamps.

LED lamps emit a relatively continuous spectrum with gaps. Critics of these lamps are hung up on this spectrum, which has a gap at 490 nm for technological reasons and ends at 420 nm, i.e., does not range into the UV portion of the spectrum. They argue that LED illumination would therefore not facilitate colour vision in the UV-A range and falsify colours for reptiles as a result. However, like with fluorescent tubes, UV halogen metal vapour lamps have to be used for the purpose of supplying UV-B light in addition so that this shortcoming should actually be properly compensated for. Detailed studies on this aspects do not exist as yet, but it is a fact that UV fluorescent tubes will emit only small amounts of UV as well, and their spectra are banded and therefore not continuous either and fraught with gaps, too (Akeret 2017).

Colour vision in the UV-A range surely plays a major role for reptiles that select their mating partners based on colour attributes, signal their readiness to mate by colour changes, or express their mood in this manner, as chameleons do, for example. Egyptian Tortoises, on the other hand, do not depend on any of these, and the absence of certain impressions of colours should do no more than falsify the chromatic appearance of feeder plants and of edible flowers in particular. The possibility cannot be entirely excluded that a terrarium illumination based exclusively on LED may change the colours of their ambience for *Testudo kleinmanni* as well. The observations on various species of reptiles mentioned earlier did not report on any negative changes in behaviour, activity or feeding after the lighting of their terraria was changed to LED, however (Kunz 2015, Schramm 2016).

A major advantage of LED lamps compared to fluorescent tubes is their effective light yield. LED lamps reach remarkable values of 150 lm/W and more, as compared to some 80 lm/W at maximum in the case of fluorescent tubes. LED lamps furthermore give off their light in a limited angle of usually between 120 and 160 degrees so that no light is lost through dispersion away from the ground, as is the case with fluorescent tubes, rendering the use of reflectors unnecessary. In fact, all their light is directed right to the bottom of the terrarium, and there it just cannot be bright enough for our Egyptian Tortoises.

Another advantage of LED lamps is their long life expectancy. Some models are supposed to last for some 50,000 hours, which means they would in theory need to be replaced only after eleven years if they are operated for twelve hours a day. A long-term trial conducted by the German consumer watchdog organisation Stiftung Warentest (2017) saw high-quality LED lamps still working without problems after more than 30,000 hours of operation, emitting light at undiminished brightness, unchanged colour rendering, and surviving roughly a million state changes of on and off.

A positive side-effect in the terrarium is seen in live plants that will thrive particularly well under the light of LED lamps (AKERET 2015, 2017, FISCHER 2017/2018). LED lamps exist in a multitude of designs, ranging from the pear-shaped globe with the traditional E14, E27 or bayonet socket via spotlights and compact lamps to tubes so that a fitting LED lamp can be found virtually for every application and every lighting effect. LED tubes make for a highly effective basic illumination of terraria, and additional LED lamps may be used to create "insolated" islands, and the entire arrangement can be timed or run via timed dimmers to create phases of dawn and dusk.

An example of an LED globe with an E27 socket for ordinary globe fittings.

Photograph Ricarda Schramm

Just one example out of many is Philips' Master LEDtube Universal UO 865 T8, which is a tube with a high light yield of 154 lm/W (UO = Ultra Output), an emission angle of 160° and an energy consumption of a mere 24 Watt in its longest model of 150 cm. Philips recently even added to their range an innovative slim T5 tube: the Master LEDtube InstantFit EVG HO 865 T5.

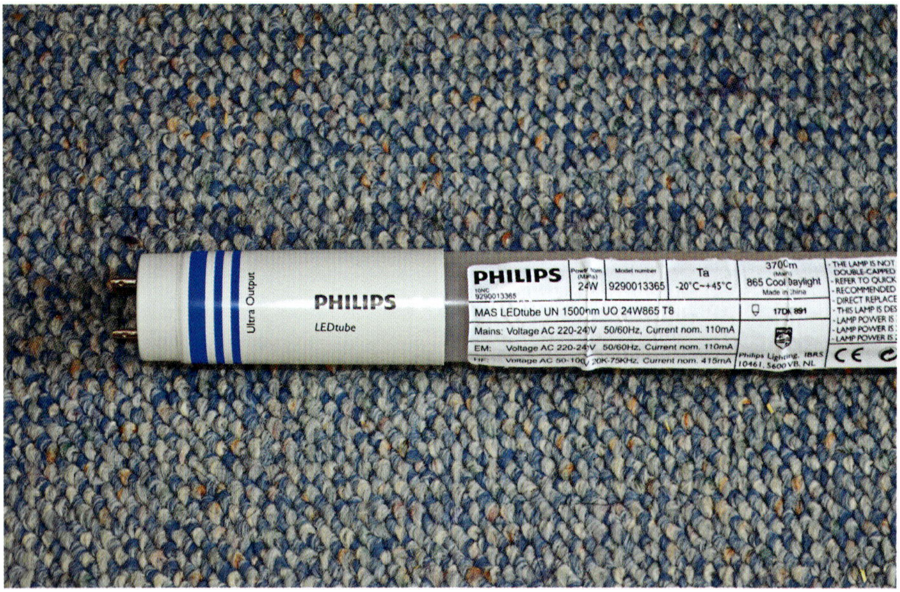

An example: Philips Master LEDtube Universal UO 865 T8 150 cm

Photograph Ricarda Schramm

All LED lamps are operated with their own ballast units. These are inserted into the operation unit in the case of LED tubes, whereas they will be integrated into the socket in globular LED lamps.

Zoo Med, under the name ReptiSun® LED UVB Terrarium Hood, offers an integrated solution composed of a ReptiSun® 5.0 T5 HO fluorescent tube and several LED lamps that may be programmed independently.

The technological development of LEDs is rapid at present, with effectiveness, colour rendering, and light colour being improved all the time. At least in theory, it is possible even now to cover the desired UV range by combining numerous LEDs with different wavelengths even if it may still be a major undertaking and rather costly at this stage (Geissel & Kober 2015, Grimm 2016).

It is but certain a field that will be thrilling to follow further.

Overview: LED lamps have an extremely effective light yield of up to 150 lm/W, do not waste light by directing it up, do not require a reflector, and are very long lived. LEDs emit a relatively continuous spectrum, which does not (as yet), however, extend into the UV range. They do not therefore facilitate colour vision in the UV-A range for reptiles. Plants do particularly well under LED lighting in the terrarium.

Basking Spots, UV Halogen Metal Vapour Lamps

General Considerations

In nature, the sun is the source of abiotic factors that are vital (not only) to reptiles, i.e., bright light, heat (infrared radiation), UV-A and UV-B, and all of these at just the right proportions. Reptiles have adapted to these conditions within their natural habitats during the courses of their evolution. They will by instinct associate spots that are brightly lit from above with solar radiation and actively seek them out for absorbing heat and UV-B radiation. Poikilothermic reptiles need to take in heat from external sources to bring their organisms to "operating temperature" on the one, and to facilitate the vital synthesis of vitamin D_3 in their skin with the aid of UV-B radiation on the other hand. Simply put, UV-B will transform provitamin D_3 in the skin into previtamin D_3 that is then transformed further into vitamin D_3 by the action of heat. This substance will be rather instable at first and therefore needs to be converted into a storable form that is available for use to the body. Amongst others, vitamin D_3 is essential to the calcium metabolism, which is the incorporation of calcium into the bones. In fact, no calcium can be deposited in the bone matrix without vitamin D_3, and no vitamin D_3 can be synthesised without UV-B and heat.

A shortage of vitamin D_3 causes rickets in juvenile and osteomalacia (soft bone syndrome) in adult reptiles, both of which are clearly seen from the outside. Substituting their diet with vitamins will be ineffective if no UV-B radiation is provided as well (Hoby et al. 2010), and cannot therefore lead to the desired results (Allen et al. 1998). Dosing such supplementation properly is difficult, however, and the required/dangerous doses have not yet been researched for many reptile species. Overdosing vitamin D_3 is known to cause problems due to its being stored in the liver tissues and may in the end even cause a catabolism ("removal") of calcium (Diegel 2010, 2011).

Based on research using human skin samples, the wavelength range from

280 nm through 305 nm is deemed effective for the synthesis of vitamin D_3 (MacLaughlin et al. 1982). Vitamin D_3 does not remain stable for long and can itself be destroyed again by UV radiation between 280 and 335 nm and turned into byproducts (Webb et al. 1989, Wunderlich 2017). The formation of the stable storage form (calcidiol) from vitamin D_3 is reduced by the action of UV radiation between 310 and 335 nm, but calcidiol itself is insensitive to UV radiation (Hollis 2011). While the mentioned wavelength ranges in part differ throughout the literature the fact will be unchallenged that the synthesis and metabolism of vitamin D_3 are balanced to such perfection that no overdosing can occur under natural sunlight. In order to provide Egyptian tortoises with an optimum supply in human care, the lamps employed for creating basking spots should emit light that it as similar to sunlight as possible and this means that their spectrum must cover the range of 280 to 335 nm as well next to providing the required heat and brightness.

All this of course raises one question: How much UV-B do reptiles need to be able to synthesise enough Vitamin D_3? This puzzle was researched by Prof. Gary Ferguson and his team between 2002 and 2008 by means of taking direct measurements from fifteen species of reptiles in their natural habitats (Ferguson et al. 2010, 2014). Their results produced four categories of reptiles that are based upon their thermoregulation habits and are now known as the Ferguson Zones:

Zone	Ecology	UV Index range of average values	UV Index maximum
1	crepuscular, living in full shade	0–0.7	0.6–1.4
2	occ. basking in the open, more comm. in semishade	0.7–1,0	1.1 – 3.0
3	occ. basking in semishade, more comm. in the open	1.0 – 2,6	2.9 – 7.4
4	heliophilic, even out in the sun around noon	2.6 – 3,5	4.5 – 9.5

Measurements were taken with various solarmeters. The defined UV Index (UVI) is actually a measurement of the risk of sunburn, with taking protective measures being recommended to humans from a UVI ≥ 3. 5–6 being considered high, and anything above 8 dangerous.

Frances BAINES et al. (2016) picked up on the results by FERGUSON et al. (loc. cit.) and worked out classifications into the four Ferguson Zones for numerous additional reptile species, partly based on measurements taken in their natural habitats and partly based on observations of their behaviours in human care. Supported by comparative measurements taken from various suitable lamps with UV emissions, they produced a guide to their selection as per the Ferguson Zones. Their list of reptile species investigated also included *Testudo kleinmanni*, which they classified in Zone 3. Most reptiles living in warm regions will avoid exposure to the midday sun, as temperatures are too high for comfort at this time and there is a distinct risk of overheating (DIEGEL 2010, FERGUSON et al. 2010, KÖHLER 2011a). This also holds true for *Testudo kleinmanni*, which will, if at all, leave their shelters only in rare instances in summer, and then do so only in the hours of early morning or late evening, but otherwise be aestivating safely sheltered underground. Egyptian Tortoises expose themselves to full sun only during their activity period from autumn through spring. Reptiles are perfectly adapted to the conditions of their habitats also in terms of exploiting the available solar radiation so that they will receive just as much UV-B radiation as they need for their vitamin D_3 synthesis to run within the optimal range and the risk of suffering damages from such exposure to be minimised. It is exactly for this reason why lamps with an output of UV radiation are used in human care that will emulate natural sunlight as closely as possible with regard to spectrum, wavelengths of UV-B emission, heat and brightness. Only if all components complement each other can reptiles continue to rely on their instincts in a terrarium as they can in the wild. Otherwise, some types of lamps will cause burns to the skin or eyes if their output of UV-B goes along with too little heat and/or brightness, suggesting to the instincts of the reptiles that it would be safe to stay exposed to their emission for what will prove to have been too long in the end, or emit UV radiation that is too short-waved (BAINES 2008, 2010, BIDMON 2011, GARDINER et al. 2009, HEIKKILÄ et al. 2009, KARSTEN et al. 2009, KÖHLER 2011a, b, WUNDERLICH 2012a, 2016, 2018).

The first choice for supplying UV radiation to reptiles kept in terraria has for long been a lamp that was originally developed by Osram for the prophylactic treatment and therapy of rickets in children, the Ultra Vitalux 300 W or the identically built Radium Sanolux 300 W. This is an emitter of mixed light with a very low light yield of ca. 20 lm/W, but which will give off very intense UV radiation with a UVI of about 10. The Osram Ultra Vitalux must therefore be operated at a large distance of at least 80–100 cm to the animals and only for brief intermittent periods of 15–30 minutes on a few days per week (DIEGEL 2010, GEISSEL & KOBER 2015, WUNDERLICH 2018). It is thus unsuited for continuous daily use at basking spots, but basically without an alternative for treating cases of rickets and osteomalacia in reptiles.

UV-halogen metal vapour lamps: Many UV lamps recommended by manufacturers in fact prove to emit UV-B in insufficient quantities, or their emissions cannot even reach the animals due to their distance to the lamp (DIEGEL 2010, WUNDERLICH 2012a, 2018). The requirements outlined above are fully met by UV halogen metal vapour lamps (UV-MH), though. These differ from ordinary halogen metal vapour lamps by a filling with different metal halogenids that will as a result produce light of a different spectrum. UV-MH emit light with a spectrum that comes very close to that of natural sunlight, has only a few gaps, and goes along with a very high degrees of brightness and thermal radiation.

2007 saw Lucky Reptile introduce to the market their Bright Sun UV that was the first halogen metal vapour lamp for terrarium purposes to give off a spectrum with a facilitating effect on the synthesis of vitamin D_3. This novelty was at the time hailed as a "revolution" and rightly so, finding use in many terrarium collections quickly. It was followed in 2011 by the "UV-MH" manufactured by the Swiss company X-Reptile, and on 26 Oct. 2013 by the release of the "Premium UVB MD-Lamp" by the Austrian Reptiles Expert! GmbH on occasion of a meeting of the Internationale Schildkröten Vereinigung (ISV; “International Chelonian Association”) in St. Pölten, Austria.

Since that time, a number of similar UV-MH lamps made by other manufacturers have appeared on the market, a complete list of which cannot be provided here for reasons of space. They have been tested and reviewed in various other publications (DIEGEL 2010, FISCHER 2017/2018, GEISSEL & KOBER 2015, GEISSEL 2016, GRIMM 2016, KÖHLER 2011a, WUNDERLICH 2017, 2018). All UV-MH lamps need to be operated with their dedicated ballast units, which will also be sized to exactly match their individual wattages.

All the above suggests that UV-MH lamps are at present the first choice when it comes to supplying terrarium animals with UV-B radiation. A tank with a floor space of 200 x 100 cm will need to be outfitted with two or three such lamps with

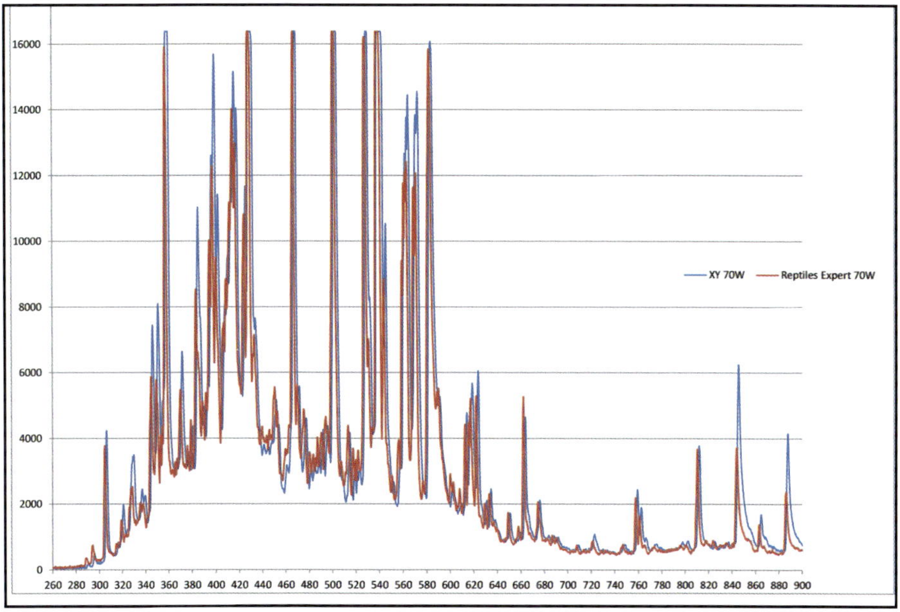

Spectra of UV-MH 70 Watt: Intensity of radiation (y axis) relative to wavelength in nm (x axis)
Top: Graphic ©Andreas Krb (Reptiles Expert! GmbH, Austria); bottom: Lucky Reptile Bright Sun UV Desert, data from package. Photograph Ricarda Schramm

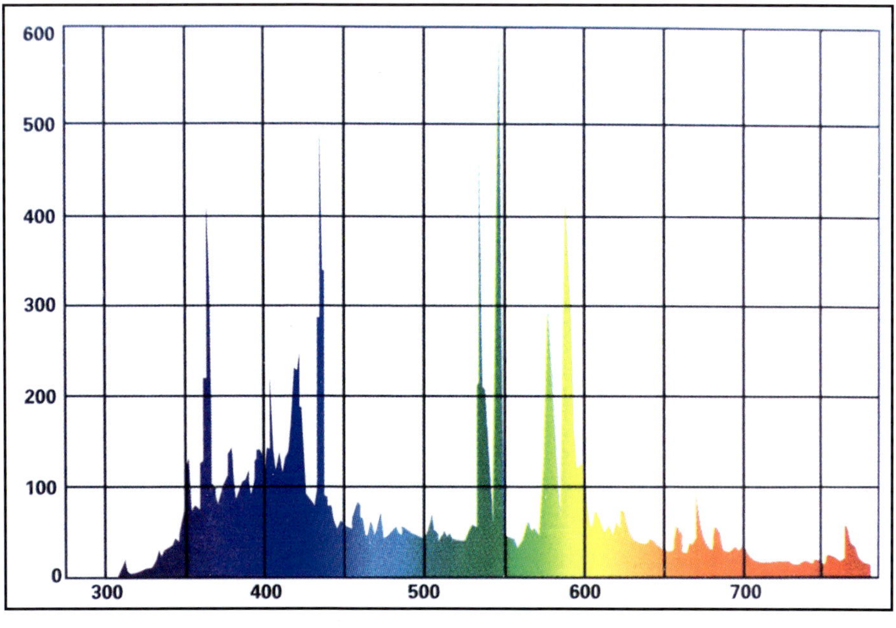

Examples of UV-MH lamps
Left: Lucky Reptile Bright Sun UV Desert 70 Watt with Osram ballast unit.
Photograph Ricarda Schramm

Right: Reptiles Expert! Premium UVB MD lamps.
Photograph Franziska Biedenweg

an output of 70 Watt each, and attention must be paid to providing enough free space and shaded situations outside of their cones of light so that the tortoises can leave their basking spots when they feel like doing so. Ingo Diegel (2011) recommended installing ordinary, UV-B-free basking spots alongside those illuminated with UV-MH lamps in order to give the animals the option to heat up without simultaneously taking in UV-B.

Egyptian Tortoises are categorised in Ferguson Zone 3, which means that they will require powerful, bright UV-B spotlights. A suitable UV-MH lamp to this effect is the "Desert" type. Ferguson et al. (2010, 2014) gave a maximum UVI of 2.9 to 7.4 for Zone 3, but theirs are maximum values that will not necessarily be reached under these lamps. Baines et al. (2016) considered a UVI of 2.6 to 3.5 sufficient at basking spots. Because the intensity of the UV-B radiation changes with the distance between the lamp and the surface of the tortoise shell, lamps of this kind will need to be set up with great care. Guidance is provided in the form of recommended distances and measurements on product packages, in user

A basking spot (UV-MH) in a terrarium,

Lucky Reptile Bright Sun UV Desert 70 Watt

Photograph Ricarda Schramm

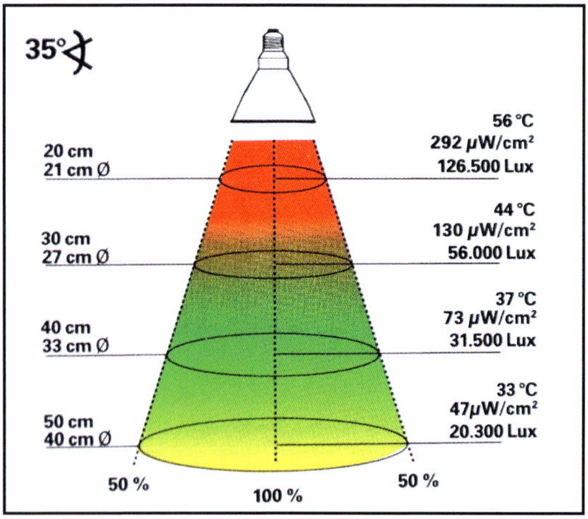

Data on temperature, intensity of UV-B radiation, and brightness relative to distance.

Lucky Reptile Bright Sun UV Desert 70 Watt Spot

Photograph Ricarda Schramm

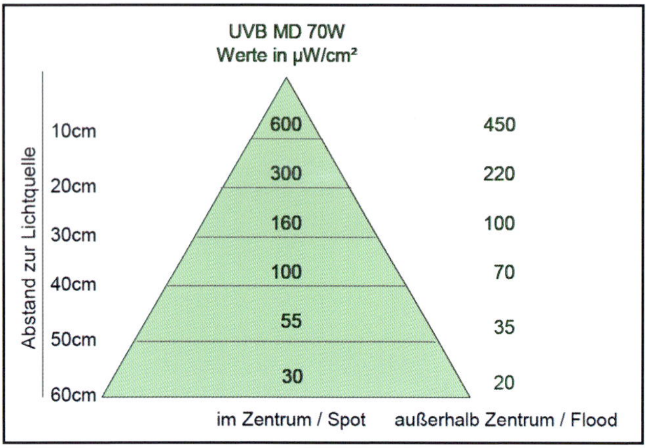

Intensity of UV-B radiation relative to distance

Reptiles Expert! Premium UVB MD lamp

Graphic: ©Andreas Krb (Reptiles Expert! GmbH)

manuals, or in the data sheets of the lamps and has to be heeded. A difficulty here is that the manufacturers' values are not given as UVI, but as the intensity of UV-B radiation in µW/cm², which cannot be readily converted into the former. This value should not exceed 150 µW/cm² (Geissel & Kober 2015). It must also be kept in mind that the UV-B output of these lamps will decrease over time and that they need to be replaced after a while. Many of the mentioned authors therefore recommend that keepers monitor this output by measuring the UVI with a solarmeter (e.g., Solarmeter 6.5 by Solartech or ZooMed), which will be the only safe way of learning when a lamp has to be replaced. The unfortunate fact here is that such a solarmeter will be rather costly. UV-MH lamps should in any case be exchanged after they will have reached their maximum life expectancy as per manufacturer's specifications, which will usually be the case after 8 to 12 months of operation.

Measuring the UVI with the Solarmeter 6.5 at a distance of about 30 cm between lamp and tortoise shell

Photograph

Franziska Biedenweg

Various authors have published overviews of the measurements they have taken under a variety of lamps. Because these values partly differ from one author to the next, and these authors themselves reported that measurements may differ from one lamp to the next even if they were of the same type, we have done without providing exact data for individual lamps here. If you are interested in this subject you will have to consult the comprehensive literature in any case.

Pointers to this effect can be derived from the temperatures that are reached beneath a UV-MH lamp, however. They should range between 35 and 40 °C, and are most simply measured with an ordinary thermometer, or with greater precision with a non-contact laser infrared thermoscanner or even with a common ear thermometer on the surface of the tortoise shell (Geissel 2016). If temperatures are too high, the tortoises will instinctively remove themselves from the illuminated spot in order to protect themselves from overheating (Köhler 2011a).

Measuring the temperature on the shell of a Testudo kleinmanni at a basking spot by means of an ordinary ear thermometer.

Photograph Ricarda Schramm

The mentioned UV-MH lamps are available from the terrarium supply trade as Flood and Spot models. These differ in their emission angles, which will be about 60 ° in the Flood variety and 35 ° in the Spot one. The radiation emitted by a Flood model is thus dispersed over a larger area whereas that of a Spot-type lamp will be more focused and means that both the intensity of the UV-B radiation and the brightness (Lux) relative to the illuminated area will be greater in the Spot variety and come closer to the value of ca. 100,000 lx of natural sunlight. This effect is actually quite readily visible to the eye in the difference in brightness below a Flood- as compared to a Spot-type lamp and also stands out in the published values for brightness and UV-B intensity (Geissel 2016, Geissel & Kober 2015, Wunderlich 2018).

The daily operating periods of the UV-MH lamps are managed by means of simple timers and adapted in duration to the changes in the hours of sunshine during the course of the year in the natural habitats of the tortoises.

Basking Testudo kleinmanni will often extend their legs in order to also take in heat and UV-B radiation via their skin.

Photograph Franziska Biedenweg

Overview: According to today's state of technology, UV-halogen metal vapour lamps are the first choice when it comes to providing tortoises with vital UV-B radiation in sufficient amounts.
These lamps need to be operated with fitting ballast units, and the dimensions of the terrarium will determine how many lamps need to be installed. They have to be mounted at the right distance to the tortoises for which manufacturer's recommendations have to be followed.
Enough space has to be provided around the cones of light of the UV lamps in order to enable the animals to retreat to unaffected areas if and when they want to.
The intensity of the UV-B radiation of all UV-MH lamps will decrease over time for which reason they have to be replaced at the latest after their life expectancy, as indicated by the manufacturer, has expired, i.e., usually after some 8 to 12 months.

Summary

The most important points of the previous paragraphs are summarized here for ease of reference.

Basic Illumination:

- As bright as possible, which means high values for Lux (lx), Lumen (lm), or Lumen per Watt (lm/W);
- Light temperature between 5,500 and 6,000 K;
- UV-A facilitates colour vision in the UV portion of the spectrum.
- The UV-B radiation emitted by UV fluorescent tubes is insufficient for the needs of the tortoises.
- Fluo tubes must cover the entire length of the terrarium.
- UV fluo tubes, halogen metal vapour lamps, and LED tubes require fitting ballast units
- UV fluo tubes, HQI spots and flooders have to be replaced after one year of operation at the latest.

Basking Spots, UV-B:

There is no alternative to installing UV metal halogen vapour lamps (UV-MH); they are at present the first choice when it comes to ensuring the tortoises can synthesise the vital vitamin D_3. UV-MH lamps provide UV-B radiation together with very bright light and heat.

- The "Desert" variety is best suited for *Testudo kleinmanni*.
- UV-MH lamps can only be operated with fitting ballast units, which are specifically designed for their individual wattages.
- The dimensions of the terrarium determine how many lamps will have to be installed.
- The lamps have to be mounted at the right distance to the tortoises for which manufacturer's recommendations have to be followed.
- Enough space has to be provided around the cones of light of the UV lamps in order to enable the animals to retreat to unaffected areas if and when they want to.
- The intensity of the UV-B radiation of all UV-MH lamps will decrease over time for which reason they have to be replaced at the latest after some 8 to 12 months.

Manufacturer/Model, e.g., Lucky Reptile Bright Sun UV Desert; Reptiles Expert! GmbH Premium UVB MD Lamp; X-Reptile UV-MH.

Tabulated Overview of Lamps for Basic Illumination:

Attributes	UV fluorescent tubes	Halogen metal vapour lamps	LED
Manufacturer Model Examples	Lucky Reptile UV Sun Zoo Med ReptiSun® Exo Terra® Repti Glo	Osram, Philips HQI, HCI, CDM	Tubes Philips e.g., Master LEDtube Universal 865 T8, Master LEDtube Instant Fit 865 T5
UV-A	**+**	**+**	**–**
Light yield	ca. 80 lm/W	ca. 117 lm/W	up to 150 lm/W **+**
Advantages	some UV-A emission	UV-A emission	highest light yield, no major thermal emission, must not be replaced before ceasing to work
Disadvantages	lower light yield, reflector required, must be replaced after ca. 1 year	high thermal emission, effective area limited, must be replaced after ca. 1 year	no UV-A emission

To make the most of the various types of lamps, you may want to consider combining, e.g., UV fluorescent tubes and LED tubes.

Basking in a state of complete relaxation

Photograph Franziska Biedenweg

Ambient Temperatures

The course of temperature plays a major role in the annual cycle of the Egyptian Tortoise, in particular with regard to the twice-yearly switch between activity and summer dormancy. This annual cycle has to be facilitated for *Testudo kleinmanni* kept in human care as well. To this end, the ambient temperatures are gradually adjusted at regular intervals on the basis of climate data from their natural habitats. Beginning with minimum values of 10–15 °C at night and 18–20 °C during the day in December and January, the values are progressively raised to a maximum of 25–26 °C at night and 35–38 °C during the day in August, only to gradually decrease once more towards December. How to achieve this is quite impossible to specify in a generalised manner and by which technological means exactly, however, as reaching the required temperatures will depend greatly on the individual situation at the spot in which a terrarium has been set up and its type. You will instead have to measure temperatures and find out which sources of heat must be switched on or off and when and which ventilation surfaces must be opened or closed to provide which results. This will obviously require you to install thermometers, preferably in all strategic places. A proven tool to this effect is the lightless Elstein ceramic heaters, which will give off only heat and no light and may therefore be used even at night. During the day, additional heaters in the shape of spotlights may be added as required, of which a multitude of suitable models are available from the terrarium supply trade.

Left: An Elstein ceramic heater; right: An example of a hotspot

Photographs Ricarda Schramm

Either type of these heaters has a regular E27 socket and does not require a ballast unit. Both must be operated with a ceramic fitting that is designed for higher wattages for safety reasons, however. A reflector may also be sensible to direct all heat down where it is needed.

Temperatures can be managed automatically by using a thermotimer, like, e.g., the TFA Dostmann, Lucky Reptile Thermo Control PRO II, or similar thermostats that offer parallel channels for timing and controlling different day and night temperatures. The two mentioned thermostats show the actual temperature on their displays. Their thermosensors should be fixed as far as possible from the source of heat and at about the level of the tortoises.

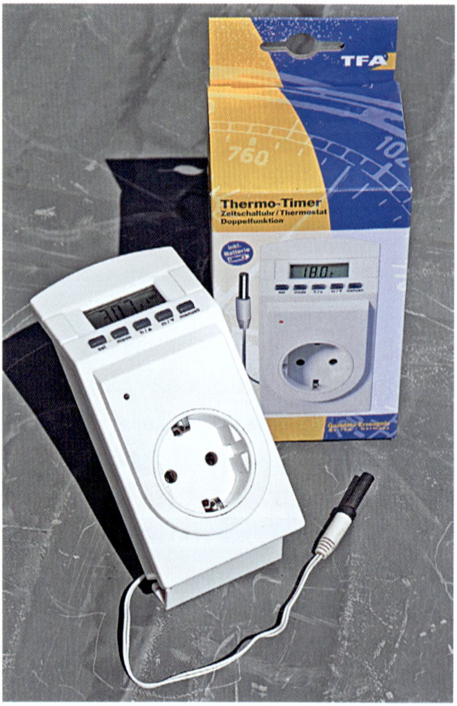

The thermotimer TFA Dostmann
Photograph Ricarda Schramm

Humidity

The natural habitats of the Egyptian Tortoise lie in a relatively narrow strip along the coast where winds from the sea carry humid air inland all year round. This results in the average degree of relative humidity being subject only to minor fluctuations (see p. 52). From about October through April, a relative humidity of 55–65 % is observed during the days and of 70–80 % at night, whereas values are marginally lower at 50–60 % during the days and 65–75 % at night from May through September.

Because no rain showers are to be emulated throughout the summer months, maintaining these elevated levels of humidity will be particularly important for the wellbeing of the tortoises.

The required humidity values inside the tortoise terrarium are most easily reached by employing an ultrasound fogger, like, e.g., the Lucky Reptile Super Fog II, which in this instance is controlled via a hygrostat like, e.g., the Lucky Reptile Humidity Control II, or corresponding models by other manufacturers. The mentioned hygrostat permits to program various timed intervals with various humidity values that should be adapted to the values known from the wild during the course of the year. The sensor of the hygrostat should be positioned as far from the outlet of the fogger as possible.

Fogger,
Lucky Reptile
Super Fog II

Hygrostat
Lucky Reptile
Humidity Control II

Photographs
Ricarda Schramm

Display of the hygrostat mounted to the outside of the terrarium

Photograph Ricarda Schramm

The sensor of the hygrostats is affixed to the wall diametrically opposite of the outlet of the fogger inside the terrarium (marked). Photograph Ricarda Schramm

According to manufacturers' instructions, foggers must be operated using only distilled water because the membranes of the pumps may otherwise become clogged with lime deposits. Considering how close the natural habitats of Egyptian Tortoises are to the sea it is quite safe to suppose that the humid air they breathe will contain salt, and salt is known to have positive effects on their respiratory system. It is for exactly this reason why some keepers will fortify the water in the fogger with some sea salt (Dworschak 2010, 2012), add it to the pressure mister (Schmidt 2009) or to the drinking water for their tortoises (Fritzsche & Fritzsche 2002). Only pure sea salt should be used for this purpose, never ordinary table salt. As to its dosage, a half to a flush teaspoon of sea salt may be added to each or only to every other filling of the fogger reservoir, but it must first be fully dissolved in water before it is added to the tank in any case.

The reservoir tank and all pipes and tubes of the fogger have to be cleaned thoroughly at regular intervals in order to forestall a proliferation and the dispersal of germs.

A fogger operating in the early morning hours.

Photograph Ricarda Schramm

Note: All technological equipment discussed in the preceding paragraphs should be installed by a professional and tested as to its operational safety.

A terrarium with live plants

Photograph Ricarda Schramm

Planting the Terrarium

Live plants in a terrarium do not only serve to enhance its appearance, but also play an important role in humidity management, for creating structure and visual barriers, and Egyptian Tortoises like to use them for sheltering. In fact, the tortoises will often bury themselves in the sand around the roots of plants.

Because the plants that are native to the natural habitats of *Testudo kleinmanni* are unavailable at our latitudes, we will have to make do with other species. Keepers of Egyptian Tortoises most commonly use succulents for this purpose, as these plants will be able to store water in their specialised organs and in this manner are adapted to surviving dry periods (Branser 2011a, b, Reinarz 2008, Scheffel 2014, Schmidt 2009). Suitable plants are found in groups such as aloes, bromeliads, inchplants, succulents, zebra plants, cacti and elephant's feet (*Beaucarnea*). Stone plants, which originally were indigenous to South Africa, but are by now distributed all over the Mediterranean and thus can be found in the distribution range of *Testudo kleinmanni*, are suitable for greening their terraria as well.

Stone plants such as Left:

Pigface Carpobrotus acinaciformis, syn. C. edulis var. rubescens

or

Hottentot fig Carpobrotus edulis (white-flowered) are encountered in the Mediterranean and can be used for planting terraria for the Egyptian Tortoise.

Grasses are another widely used option. Some of the plants may simultaneously be made use of as feeder plants, which is an aspect that will be discussed in the chapter Diet below.

Bromeliads, and in particular the epiphytic species, originate for the largest part from subtropical rainforest regions and are therefore adapted to a climate zone with entirely different conditions than those in the habitats of *Testudo kleinmanni*. Some species have proven to do surprisingly well in desert terraria notwithstanding, and these are predominantly found amongst the non-epiphytic, terrestrial bromeliads that are robust enough to withstand dry seasons and exposure to bright sunlight.

Beat Akeret (2015) described in his book a wide selection of plants that are suitable for use in various types of terraria and provided a multitude of guidelines and tricks for how to best employ and care for them.

What you should know: As a matter of principle, only non-poisonous plants should be used for planting terraria, because the possibility cannot be excluded that the tortoises will experimentally nibble at them.

Live plants may be used in a terrarium in their individual pots or planted right into the bottom substrate. Those that will simultaneously be exploited as feeder plants are better left in their pots so that replacing them at intervals will be easier. If the tortoises are to be enabled to bury themselves amongst plant roots, they will obviously have to be planted right into the bottom substrate.

Losses must be expected to occur even in the case of the most robust plants after some time, meaning that individual plants may need to be replaced with new ones from time to time. You will have to experiment a bit and make your own experiences to find out which plants exactly meet your expectations as to their durability. However, even dead and withered plants can be left in the terrarium as they will continue to provide the tortoises with popular retreats.

Tables on the following pages:
Examples of Suitable Terrarium Plants
From top to bottom:
Photographs: Nos. 3, 6, 12 Franziska Biedenweg;
all other photos Ricarda Schramm

Family	Species	Photograph
Spiderworts Commelinaceae	*Callisia repens* Creeping Inchplant Golliwoog®	
Bromeliads Bromeliaceae	*Bilbergia nutans* Queen's Tears	
Bromeliads Bromeliaceae	*Tillandsia cyanea* Pink Quill	
Stonecrops Crassulaceae	*Crassula ovata* Lucky Plant or Money Tree	
Stonecrops Crassulaceae	*Crassula ovata* Gollum Lucky Plant or Spoon Tree	
Grass Trees Xanthorrhoeaceae Subfamily Aloes Asphodeloideae	*Aloe spp.*	

Family	Species	Photograph
Grass Trees Xanthorrhoeaceae Subfamily Aloes Asphodeloideae	*Haworthia* *(Haworthiopsis)* *attenuata* Zebra Plant	
Grass Trees Xanthorrhoeaceae Subfamily Aloes Asphodeloideae	*Haworthia* *(Haworthiopsis)* *limifolia striata* File-leafed Haworthia	
Cacti Cactaceae	*Schlumbergera* spp. Christmas Cactus	
Sedges Cyperaceae	*Uncinia rubra* Devil's grass	
True Grasses Poaceae	*Festuca cinerea* Blue Fescue	
Asparagi Asparagaceae	*Beaucarnea recurvata* syn. *Nolina recurvata* Elephant's Foot	

Plants in the terrarium are sniffed at and some may then be eaten.

Photographs: Top: Ricarda Schramm bottom: Franziska Biedenweg

Withered plants still provide shelter.

Photographs Michael Hebbeler

During winter, when *Testudo kleinmanni* is active, the plants in the terrarium will receive enough water from the emulated rain showers, but not so during summer when no rain whatsoever falls. If you were to continue watering the plants now, you would effectively disturb the aestivating Egyptian Tortoises. A solution to this conflict of interest during the hot months of summer exists in the shape of ceramic self-watering spikes that will simultaneously take care of keeping the lower layers of the bottom substrate moist. The latter is essential for creating and maintaining a favourable microclimate in the aestivation shelters that will protect the tortoises from dehydrating. Ceramic self-watering spikes for vacation plant watering are available in various sizes from a range of manufacturers. They typically come with plastic caps and thin hoses that are meant to connect them to a water reservoir for a continuous water supply. For our purposes, however, the caps are removed and the spikes buried in the bottom substrate amongst the roots of the plants. They are then very carefully filled with water at regular intervals so that they can release it slowly into their surroundings.

Ceramic watering spikes with plastic caps and hoses

Photograph Ricarda Schramm

Some additional self-watering spikes may be buried near the preferred tortoise shelters so that they can provide the required substrate moisture at the deeper levels.

Ceramic self-watering spikes buried amongst the plant roots

Photograph Ricarda Schramm

Plants provide structure to a terrarium and offer opportunities for sheltering.

Photograph Franziska Biedenweg

Structuring and Shelters

A well-structured layout of the terrarium or open-top indoor pen contributes greatly to the wellbeing of Egyptian Tortoises. It creates visual barriers and obstacles that will enable them to avoid other members of their colony, and they can find retreats and shelters that afford them a sense of safety. Structural elements will also form “landmarks” that help them to find their way around. Exhibiting a great degree of site fidelity also in their natural habitats, *Testudo kleinmanni* memorise components of their environment to orientate themselves and find back to oviposition spots, places at which fresh feeder plants are to be had, or shelters, for example.

Structural elements of this kind may be plants, rocks, pieces of roots, or little hills. They can be used to create separated areas, visual barriers, protected niches, oviposition spots, or opportunities for climbing. Your imaginativeness is basically the only limit, as long as you ensure that climbing spots are not too high or too steep and, in particular, are not situated too near to hotspots. You certainly would not want your precious tortoises to fall from great heights, land on their backs, and in the worst case scenario lie in this position right under a heat lamp for an extended period of time.

Creating structure with plants, rocks and roots.

Photograph Franziska Biedenweg

Using plants and roots to create structure and shelters.

Photographs Ricarda Schramm

Top: A cave made from a rock flake.
Bottom: Fragments of fired-clay drainage pipes can be split in halves and used as retreats.
Photographs: Top: Michael Hebbeler; bottom: Ricarda Schramm

Top: A deep cave made from several pieces of fired-clay drainage pipe and covered with bottom substrate.
Bottom: A view into the cave with the resident Testudo kleinmanni.
Photographs Franziska Biedenweg

Shelters and retreats can be fashioned from a wide variety of objects. Aside from rocks bridged with a rock flake or ready-to-use caves from the terrarium supply trade, elements made of unglazed fired clay such as halved flowerpots, bowls, terracotta wine rack tiles, or drainage pipes split in half, lend themselves for use as tortoise shelters.

Unglazed fired clay has the capability of absorbing water and releasing it slowly into its immediate surroundings, creating a moist and therefore cooler microclimate in the shelter. This effect is particularly expressed if the clay elements are buried so deeply in the bottom substrate that their access openings are below the surface and the clay can absorb moisture at the deepest levels for transporting it upwards. Plants that are placed in the soil right next to tortoise shelters help to keep the deeper layers of the substrate moist. They are watered by means of ceramic self-watering spikes (see p. 119) even during the dry summer season; in fact, their survival will depend on this type of water supply during this rainless period while the tortoises aestivate in their shelters. The described setup very closely emulates the situation Egyptian Tortoises exploit in the wild when they retreat into rodent burrows for the hot months of summer.

A shelter made from a halved, unglazed clay flowerpot: deeply embedded in the bottom substrate next to a plant, it will provide a moist, cool microclimate.

Photograph Ricarda Schramm

Tight, low-ceilinged retreats are given preference, even it if means crowded conditions.

Photographs: Top: Ricarda Schramm bottom: Franziska Biedenweg

Like many other species of *Testudo*, Egyptian Tortoises prefer to shelter in tight and low spaces where their dorsal shells will be in constant contact with the ceiling, as this apparently affords them with a feeling of enhanced safety.

During their activity period from October through May, *Testudo kleinmanni* living in their natural habitats seek out hiding places that differ in character from those used for the purpose of aestivation (Geffen & Mendelssohn 1989, Schneider & Schneider 2006). This behaviour is also apparent in Egyptian Tortoises kept in human care. If the ambient temperatures are moderate, they will take shelter under roots or bury themselves in the soil between the roots of live plants for spending the night. The hot summer months, on the other hand, will see them retreating mostly to deeper and therefore moister underground places. However, individual *Testudo kleinmanni* may make use of various shelters every now and then throughout the year.

This young Testudo kleinmanni has chosen to take refuge under a root.

Photograph Ricarda Schramm

Hatchlings tend to preferably conceal themselves by burying themselves right in the bottom substrate, usually at the base of a plant or under a piece of root. Only once they have grown a little older will they make use of the cavities provided as well.

A hatchling Testudo kleinmanni burying itself in the moist bottom substrate.

Photographs Franziska Biedenweg

After they have been turned into a shelter fired-clay elements can be covered with substrate on the surface. This will create not only a naturalistic appearance, but also does not subtract from the usable floor space because the resultant low "hill" is no obstacle to the agile tortoises.

A halved fired-clay flowerpot covered with bottom substrate now forms a hill that is easily overcome and does not therefore subtract from the available floor space.

Photograph Franziska Biedenweg

Shelters made from cork or willow bark are lightweight and cannot be firmly anchored in the ground so that adult tortoises will shift them from their intended spots again and again. They may therefore be fit only for hatchlings and small juveniles.

All structuring and decoration must in the end leave enough open space for the active *Testudo kleinmanni* to move about unimpeded. In fact, the open floor space should account for the largest part of the ground.

The day starts badly if the exit is blocked...

Photograph Franziska Biedenweg

Examples

Example 1 Franziska Biedenweg

The terrarium for Egyptian Tortoises

Photograph Franziska Biedenweg

This terrarium is home to three adult *Testudo kleinmanni*, one male and two females, as well as two semiadult specimens.

It is a DIY construction made of 16-mm OSB that has been impregnated against moisture on the inside with epoxy resin and measures 200 x 100 x 60 cm. It is a partly open-top design that allows for opening or closing parts of the cover for temperature management. The terrarium serves as a room partition and is therefore outfitted with glass sliding doors on both long sides. The 4 mm glass panes were cut to size by a glassier and outfitted with grip recesses and ventilation holes at the same time.

The bottom substrate is a mix of playground sand and loam at a ratio of 5:1. One corner of the tank sports a small section that is cordoned off with stones and filled with garden soil. It is used for planting or seeding fresh feeder plants, which the tortoises love to help themselves to.

Top: Recessed grip and three ventilation holes in a glass sliding door

Bottom: A cordoned-off area with garden soil for growing fresh feeder plants

Photographs Franziska Biedenweg

The feeder plant bed is happily exploited.

Photograph Franziska Biedenweg

Technological outfitting: Basic lighting of this tank is provided by four Zoo Med Nature Sun 30 W T8 fluorescent tubes of 90 cm cm in length each, two of which are mounted behind each other along the front and the other two parallel in the rear attached to the lid of the terrarium. They are run with the operating units/fittings Exo Terra Light Unit for two T8 30 W and Hagen Glomat 2 for two T8 30 W. All fluo tubes are furnished with reflectors.

Three UV metal halogen vapour lamps (UV-MH) create basking spots, and one Lucky Reptile Bright Sun UV Desert 70 W Flood and two Reptiles Expert! Premium UVB MD lamp 70 W Floodare operated via Lucky Reptile ballast units, in one instance for 70 W and in two for 35/70 W switchable.

Two Elstein lightless heat lamps can be added for creating the necessary high temperatures in summer.

A fogger, Dragon Magic Fog, ensures the required high levels of relative humidity. Its water is not fortified with salt. Rain showers during the activity period of the tortoises are emulated manually by means of a pressure mister.

The hygrostat Humidity Control Pro II by Lucky Reptile automatically manages the Exo Terra Light Unit and the fogger. This device has two integrated circuits, with one connecting to the fogger and the other serving as a timer. The second operation unit, a Hagen Glomat 2, is controlled via a digital timer. The thermostat Thermo Control Pro II by Lucky Reptile likewise offers two separate circuits (timer and thermostat) and is used to manage the two outer UV-MH lamps and a single Elstein heat lamp. The UV-MH lamp in the centre is controlled via a timer.

All control units are adjusted according to the natural climatic fluctuations, as described in the chapter *The Course of the Year, Annual Schedule* (p. 182). The fluorescent tubes are switched on and off in a staggered sequence, creating transitional periods of dawn and dusk, respectively.

Arrangement of the terrarium lamps

Photograph Franziska Biedenweg

Nursery tank: Varying numbers of hatchlings are housed in a closed-top nursery terrarium built from OSB and measuring 120 x 60 x 60 cm. It is equipped with glass sliding doors in the front and wiremesh-covered ventilation surfaces in the side walls. Its is set up in a manner very similar to the tank for the adult tortoises. Basic lighting is provided by two Osram Lumilux 54 W T5 fluorescent tubes of 114.9 cm in length, with reflectors. Two Reptiles Expert! Premium UVB MD lamps of 35 and 50 W, respectively, illuminate the basking spots. A fogger, Super Fog II by Lucky Reptile, and a management unit in the shape of a digital

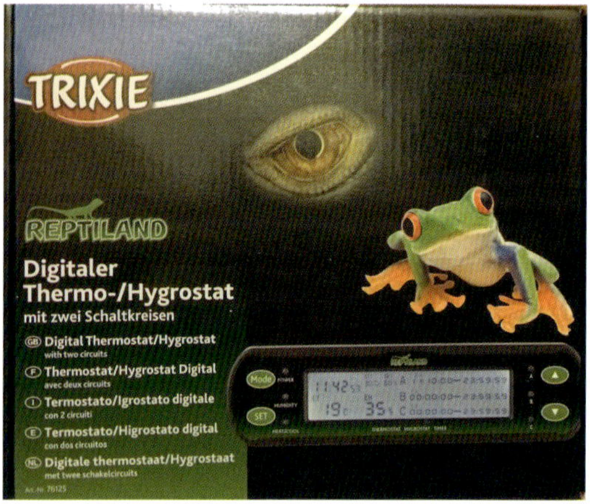

Top: A digital thermo/hygrostat with three separate circuits

Bottom: The nursery terrarium

Photographs Franziska Biedenweg

thermo/hygrostat by Trixie complete its technological outfitting. The thermo/hygrostat has three separate circuits for thermostat, timer and hygrostat, each of which offers a choice between several timed intervals and thus facilitates complete control with just one device.

Operating the small, closed-top terrarium in praxis has demonstrated that it will heat up rapidly to temperatures beyond tolerable limits in the months of autumn, winter and spring, while opening one of the sliding doors for cooling will see the levels of relative humidity drop too much. An opening in the cover lid may provide a solution to this problem.

Example 2 Ricarda Schramm

The terrarium for Egyptian Tortoises

Photograph Ricarda Schramm

The closed-top terrarium measures 200 x 100 x 100 cm and provides an artificial habitat to five semiadult *Testudo kleinmanni*.

DIY-built, the bottom, rear wall, the left side wall, and the cover consist of 18-mm OSB, while 18-mm glued laminated timber has been selected for the front and right side wall for aesthetic reasons. The cover plate is recessed down by 10 cm from the top edge of the front in order to conceal most of the electrical and electronical installations behind the ledge created thus. The front consists of two sliding doors made of 4-mm single-pane toughened glass. Ventilation is facilitated by circular grilles in the lid and low in the right side wall.

The front face of the terrarium is furthermore equipped with two hinged Plexiglas flaps of 25 cm cm in height framed with aluminium profiles. These can be flipped up and arrested in this position as to provide a safe barrier that will keep the tortoises from escaping and falling when the glass sliding doors are open for maintenance work in their terrarium or for balancing the temperature.

Note the recessed cover of the terrarium

Photograph Ricarda Schramm

Circular ventilation grille in the lid of the terrarium

Photograph Ricarda Schramm

Hinged Plexiglas flaps for securing the tank when the glass sliding doors are open

Photograph Ricarda Schramm

The bottom substrate has been mixed from building sand and loam (see p. 74), but consists only of building sand without a loam component in the cavities created from halved fired-clay flowerpots. This enables the Egyptian Tortoises to bury themselves in the loose, moist substrate with greater ease.

Two Testudo kleinmanni have buried themselves in the loose, moist bottom substrate under a halved fired-clay flowerpot.

Photograph Ricarda Schramm

Technological outfitting: Basic lighting is provided by one Philips Master LEDtube Universal UO 865 T8, 150 cm of 24 Watt, run in a operating unit/fitting Osram SubMARINE, and two large, pear-shaped LED lamps in E27 ceramic fittings. Starting with the two LED globes, the lamps are timed sequentially to come on and switch off in reverse sequence at intervals of 30 minutes in order to create a semblance of dawn and dusk, respectively. Their hours of operation are adjusted to correspond to the fluctuations of the hours of daylight during the course of the year following climate data from the wild.

Basking spots are available under two Lucky Reptile Bright Sun UV Desert

Emulated dawn by having the LED lamps
come on sequentially at intervals of 30 minutes

Photographs Ricarda Schramm

70 Watt Spots with Osram ballast units whose hours of operation are automatically timed in accordance with the natural hours of sunshine in the course of the year.

When required for creating the necessary high summer temperatures, a lightless 60 Watt Elstein heater in a ceramic fitting with a maximum capacity of 100 W and equipped with a Zoo Med Mini Deep Dome reflector is switched on in addition and controlled via a thermotimer. This combination can double up as an alternative hotspot during the activity period of the tortoises, e.g., for heating an oviposition site.

A fogger, Lucky Reptile Super Fog II, run via a hygrostat Humidity Control II by Lucky Reptile, takes care of providing the necessary level of relative humidity. Its reservoir tank holds 2.1 l of distilled water to which will be added one teaspoon of pure sea salt (thoroughly dissolved before) to every other filling. Rain showers during the activity season are additionally emulated by means of a 5 l Gloria pressure mister.

Aside from ambient temperatures and relative humidity being measured and monitored throughout the terrarium, a digital thermo/hygrometer by Terra Exotica equipped with two remote sensors displays corresponding data from a shelter. This measuring spot was made provision for by routing the cables of the sensors through the hole in the bottom of a horizontally installed fired-clay flowerpot and running them, protected behind rocks, to the measuring unit.

The sensors of the thermo/hygrometer monitor the conditions in a shelter

Photographs Ricarda Schramm

Right:
A digital thermo/hygrometer is mounted inside the terrarium

Bottom:
Partial view of the terrarium

Photographs Ricarda Schramm

Example 3: An Indoor Pen in the Attic

A particularly naturalistic husbandry strategy results from setting up a pen in a room, like in an attic below large skylights, in a conservatory, or in a greenhouse rather than using a terrarium. One major advantage of this approach is that the active Egyptian Tortoises will find much more space for walking about, and another is that the daylight filtering in from the outside will create a certain degree of basic illumination that would otherwise have to be provided with lamps. When the sun is out, the light will be brighter than anything an artificial source of light could create, there will be the associated heat on the ground, and the tortoises will visibly prefer this situation to any artificial alternative. The required high temperatures of summer can also be reached with greater ease, and ensuring the low values in winter will not be a problem either. If the conservatory or greenhouse has furthermore been roofed with Alltop® Plexiglas, the tortoises will even be able to enjoy the natural UV radiation of sunlight (Dworschak 2010, 2012, 2013, Reinarz 2008) It is thus probably the closest captive husbandry conditions can come to nature. Only on overcast days and for extending the hours of natural daylight will it be necessary to supply light from artificial sources, because the days are shorter in Central Europe than at the latitudes at which

An open pen for Testudo kleinmanni in an attic

Photograph Michael Hebbeler

Testudo kleinmanni will prefer sunlight falling into their pen through skylights to any artificial source of light.

Photographs Michael Hebbeler

the natural habitats of *Testudo kleinmanni* are situated.

It will of course still be necessary to provide basking spots under heat or UV metal halogen lamps in order to not only enable the tortoises to thermoregulate properly, but also for exposing them to vital UV-B radiation, because window glass is permeable only to UV-A light.

Managing the required elevated level of humidity may turn out a little more difficult in such an open-top, large pen in an even larger room than in a terrarium that is more or less closed all around, but it is entirely possible by means of foggers. An advantage of a large pen within an even larger room is clearly that climate parameters tend to change very gradually and therefore in a more natural manner.

The pen portrayed here is set up in an attic room of some 31 m² that is subdivided into several sections. One of the latter is home to six adult *Testudo kleinmanni*, one male and five females, in an enclosure of about 17 m² in floor space. It is outfitted with a bottom substrate mixed from building sand and loam.

The room temperature is a function of solar radiation falling in through the skylights, a large number of heat and UV lamps, and, if and when required, the effects of a central heating system for the entire room.

***Testudo kleinmanni* in a large open indoor pen**

Photograph Michael Hebbeler

Wild herbs make for optimal food for Testudo kleinmanni

Photograph Ricarda Schramm

Feeding

Wild Herbs

In our quest to provide our Egyptian Tortoises with an ambience that comes as close to nature as possible we will always have to take all relevant cues from nature. The animals have perfectly adapted to the conditions in their natural habitats during the course of their long evolutionary history so that they can now survive in a situation that must appear extremely hostile to us. Within the total concept of a naturalistic husbandry of *Testudo kleinmanni*, their diet is an aspect of major importance.

Like many other tortoise species, Egyptian Tortoises are pure vegetarians. In the wild, they will scour large tracts of land in search of edible plants that vary widely both as to their species and their constituents. They will rarely consume a whole plant, but rather nibble at certain parts. This variety in the end ensures that they will take in everything their organism needs at a balanced and therefore optimal ratio. Feeding them in human care should strive to achieve exactly this goal as well by providing a variety that will do the same for them.

Wild herbs are right at the top of the menu of *Testudo kleinmanni*. Sure, we cannot provide them with the exact same species they would find and feed on in the wild, but our latitudes host a wide variety of feeder plants that will be sufficiently similar. What is important here is they will have a very low protein content, a high content of raw fibres, and a calcium : phosphorus ratio that is significantly biased towards the former, and so match the analyses of their feeder plants in their natural habitats (Rank & Kaufman 2012). Aside from whole plants, the flowers, which are rich in minerals, are very popular treats. Native feeder plants include, amongst others, stinging nettle, cat's ear, sow thistles, hawkweeds, plantains, mulleins, evening-primroses, malvas, hawksbeards, and nippleworts, but listing all suitable wild herbs here would far exceed the scope of the present book. Detailed discussions of these, in particular for tortoises, and their constituents are provided in publications dedicated to this subject (Dennert 2008, Schramm 2017, Wegehaupt 2019 a, b).

Feeding Mediterranean tortoises that are active in summer is thus a relatively straightforward affair. Providing Egyptian Tortoises with their activity period in winter with fresh wild herbs, on the other hand, is much more of a problem. It can be resolved to some extent, however.

Wild herbs and flowers are at the very top of the menu.

Photographs: Top: Franziska Biedenweg bottom: Ricarda Schramm

Testudo kleinmanni feeding on malva leaves

Photograph Franziska Biedenweg

Preparing for Winter

It will be in late summer and autumn when you have to start making provisions for an adequate supply of fresh wild herbs in winter, which is when your *Testudo kleinmanni* are active. One option to this end is to seed wild herbs in trays that are then left to sprout at a window inside the house. This is best done at intervals so that new plants can start appearing across a longer period of time. Your harvest will then consist of both fresh shoots that contain lots of vitamins, and plants at a more advance stage of growth that can serve as first feeds for your Egyptian Tortoises when they emerge from their aestivation period in autumn.

What you should know: Only those seeds and seed mixes that are specifically made for consumption by tortoises are suitable. Mixes from the nursery or supermarket will often include plant species that are unsuitable or even poisonous.

Seeding wild herbs and letting them sprout indoors must begin already in late summer.

Photograph Ricarda Schramm

Another possibility is to dig up strong wild herbs in the wild in autumn, transplant them to flowertrays, and letting them continue to grow in a warm spot at a window indoors. They will enable you to cut off leaves for feeding your tortoises for an extended period into a season where little is available outdoors.

What you should know: Seeding and transplanting wild herbs should be done using only unfertilised garden or germinating soil, because wild plants will not tolerate fertilisers.

Digging up feeder plants and transplanting them into flowertrays in autumn, in this instance cat's ear, dandelion, hawksbeard, ribwort plantain, and wild strawberry.

Photograph Ricarda Schramm

Many feeder plants are available until far into the winter season and can readily be found in spots where there will be little or no snow. Not all of them will perish in autumn or reduce themselves fully to their roots in winter. In fact, many annual wild herbs germinate already in late summer or autumn and spend the winter as baby plants. For their part, some perennial species will maintain their above-surface parts through the winter. Larger amounts of collected feeder plants can be stored and kept fresh for several days, for which purpose they are best placed in an airy sieve that is kept in a freezer bag in a cool place.

The yield of a winter walk:
cat's ear, sow thistle, dandelion, hawksbeard, ribwort plantain, and wild strawberry.

Photograph Ricarda Schramm

Dried Herbs and Flowers

Aside from fresh feeder plants, dried herbs and flowers are very much part of the natural menu of *Testudo kleinmanni*.

Plants for drying need to be collected already in summer and then dried in a dehydrator, at minimum temperature in the oven, or simply on a grille outside in full sun. Dehydrators of all sorts and descriptions are available from commercial sources.

The dried feeder plants must then be stored in a well-aired, cool and dark place and in paper or fabric bags to prevent them from growing mouldy. Here it will be important to assure yourself that the plants have really perfectly dried through and through, which will be the case if they can be pulverised under crackling sounds between the fingers.

Not all wild herbs and flowers can be dried thus, however, with some being just too juicy for this purpose. Guidelines for which plants are suitable and which are not can be found in subject literature (Schramm 2017).

If you are for whatever reasons unable to gather your own stock of dried feeder plants, you may want to resort to what specialist turtle shops have to offer in hay for tortoises.

Dried feeder plants and flowers should be available for consumption to Egyptian Tortoises all year round and be offered even when fresh food is made available during their activity period.

A commercially manufactured dehydrator

Dried herbs and flowers

Photographs Ricarda Schramm

Dried herbs and flowers form part of the menu of Testudo kleinmanni

Photographs Ricarda Schramm

Leftovers of a meal of fresh wild herbs need not necessarily be removed the next day, but can be left to dry in the terrarium if they are not soiled with faeces or urine.

Adult *Testudo kleinmanni* may also go a day in between without fresh food during their activity period and be offered only dry herbs. Juveniles, on the other hand, should have fresh feeder plants available every day.

A juvenile Testudo kleinmanni with dried herbs and flowers

Photograph Ricarda Schramm

Cultivated Plants

If no fresh herbs are available, using cultivated plants instead may be the only alternative left. Lettuces are principally rich in protein and poor in their fibre contents and therefore actually the exact opposite of what tortoises should be fed with. It is for this reason why more hard-leafed types with their comparatively higher fibre content should be given preference, e.g., Catalogna ("giant dandelion"), endives, field lettuce from outdoor farming, radicchio, Roman and rocket. For their part, kitchen and iceberg, Lollo rosso or similar lettuces are entirely unsuited because they will be particularly rich in water and protein and poor in fibre. If you have to use lettuce to bridge a period of shortage, remember that variety will be better than one-sidedness also here.

One option to improve the fibre content of lettuces at least a little is by supplementing them with Pre Alpin Testudo made by the company Agrobs GmbH. The variety Pre Alpin Testudo Baby is to be used preferably for juveniles due to its being very finely grated, whereas Testudo Fibre will be fit for adult *Testudo kleinmanni*. The dry fibres are simply sprinkled over the lettuce and moistened so that they will be consumed right with it. It is also a good idea to offer Pre Alpin Testudo exclusively on some days in between, but it must always be moistened. Not all *Testudo kleinmanni* are willing to take in the dry fibres without at least a little lettuce, however.

Adding Pre Alpin Testudo Baby will improve the fibre content.

Photograph Ricarda Schramm

Egyptian Tortoises must never be fed with tomatoes, fruit or vegetables. Tomatoes strip the organism of calcium, the fruit sugar of fruit favours the proliferation of intestinal parasites, and vegetables are particularly rich in proteins. If anything, carrots and zucchini may be added as very rare treats.

Field lettuce from the veggie patch
makes for a popular treat amongst *Testudo kleinmanni.*

Photograph Ricarda Schramm

Adding Pre Alpin Testudo Baby will improve the fibre content of lettuces if the latter has to be used for feeding Testudo kleinmanni

Photographs Ricarda Schramm

Finely or coarsely grated, carrots are a controversial food item. Some Egyptian Tortoises will reject them completely, whereas others may fall over themselves to eat them. You will have to try and find out how your specimens are inclined.

Widely popular, on the other hand, are Christmas cacti (*Schlumbergera* spp.) and Golliwoog® or creeping inchplant (*Callisia repens)* that have already been mentioned as plants for the terrarium above. Both clearly enrich the menu for *Testudo kleinmanni*, and both the flowers and the leaf segments will be consumed with visible appetite. Christmas cacti unfold their flowers right in the time when they will be most needed, i.e., in winter, and can then be used as fresh feed.

The two mentioned plants can be established in the terrarium right with their pots so that they will serve both as sensible items of decoration and as food that is not available from a feeding tray, but, like in the wild, has to be located and accessed first.

What you should know: Newly acquired plants are best replanted immediately into larger pots filled with seedling mix soil, because most of them will have filled their pots completely with their roots and not grow much further.

Left: Creeping inchplant (Callisia repens) or Golliwoog®

Right: Christmas cactus (Schlumbergera sp.).

Photographs Ricarda Schramm

Testudo kleinmanni have a taste for both the leaf segments and the flowers of Christmas cacti.

Photographs Franziska Biedenweg

The tender leaflets of Golliwoog® are right to the taste of Testudo kleinmanni and worth the effort of climbing.

Photographs: Top: Ricarda Schramm bottom: Franziska Biedenweg

Food Supplements

Calcium

Providing captive-kept tortoises with calcium in sufficient amounts is essential. Like all vertebrates, they need calcium for the formation and maintenance of their skeleton. Calcium is but also involved in a multitude of metabolic processes like the transmission of nervous signals, the activation of enzymes, the coagulation of blood, and in muscle activity. These metabolic functions take priority over its use for bone formation when it comes to the distribution of ingested calcium throughout the body. In the cases of an undersupply with calcium or a calcium deficiency, this means that calcium may even be mobilised from the bone tissue in order to keep these vital metabolic processes running.

In tortoises, the proportion of the skeleton in the total body weight is very much larger than in other vertebrates, with the relative weight of the shell accounting for up to 35% of the body weight (BW) in adult tortoises and for some 21%BW in hatchlings (KOPSCH 2006). These figures demonstrate that the relative weight of the shell will increase from hatchling to adult age as a result of bone matter being added to it through growth. Tortoises will grow throughout life, even if increasingly slowly after becoming adult. For comparison, in birds, the relative weight of the skeleton makes up some 8–10%BW, and in humans about 10–15%BW. Growing more rapidly, hatchlings and juveniles require particularly large amounts of both calcium and phosphorus for growing and solidifying their bones and the shell. If their diet is deficient in calcium, rickets will be the inevitable consequence in juveniles and osteomalacia in adults, which is then most clearly visible in a soft and deformed shell. Adult females also have a greater need for calcium when they have to shell their eggs prior to depositing them. The nutritional content of the eggshell of *Testudo kleinmanni*, measured in the dry matter, will therefore, unsurprisingly, be made up for a large part of calcium (HARTMANN 2009).

The successful incorporation of calcium into the bone matter depends to a deciding extent on a sufficient supply with UV-B radiation without which the synthesis of vitamin D_3 would not be possible (see p. 93), and without UV-B and vitamin D_3 even the most optimal supply with calcium will do nothing for a tortoise.

In the wild, *Testudo kleinmanni* cover their need for calcium for the largest part by consuming feeder plants whose calcium : phosphorus ratios are highly biased towards the former (RANK & KAUFMAN 2012). This calcium : phosphorus ratio (Ca:P) therefore warrants a closer look.

Phosphorus, or phosphate, is likewise required by the organism to facilitate important metabolic processes. Phosphorus is a constituent of the DNA, of certain proteins and enzymes, plays a role in the storage of energy, and plays a key role in the formation of bone matter. Portrayed in a simplified manner, an excess of phosphorus will mobilise and strip the bones of their calcium contents, however. It is therefore a logical consequence that these two elements have to be present at just the right proportion, which means there must be more calcium than phosphorus.

Feeder plants with a high Ca:P ratio include, e.g., broadleaf plantain, red clover, dandelion, lucerne and malvae. Lettuces and vegetables, on the other hand and with few exceptions, all have a very low calcium content and therefore an unfavourable calcium : phosphorus ratio and will thus cause all the known and dreaded problems if they are used without added calcium for feeding tortoises over extended periods of time (Dennert 2008, Kölle 2009).

Nobody will of course take the pains of analysing the Ca : P ratio of each and every feeder plant and select them only as per these results. What becomes clear, however, is that the dangers of an undersupply with calcium or an unfavourable Ca : P ratio lurk right in the food your tortoises are supplied with and need to be eliminated by your making available additional calcium. Only by doing so will you be able to forestall calcium shortages or counteract the effects of unfavourable Ca : P ratios.

In a captive setting, it is fairly safe to suppose that an appropriate supply with calcium cannot be ensured via the feeder plants alone. Calcium therefore will have to be made available in the shape of other sources in addition. It must never be sprinkled over food, however, but rather be freely accessible so that the tortoises can help themselves to it if and when they feel they need to do so. They are the only ones that will know for sure, and overdosing calcium is entirely possible.

Cuttlefish shell: The most commonly exploited source of calcium for tortoises is cuttlefish shell, the internal skeletons of squids that consist of calcium carbonate. They are made freely available to the tortoises whole, in fragments, or crushed. When buying cuttlefish shell you will need to assure yourself that you pick the untreated type, because some are fortified with vitamins for birds.

It is a common observation that cuttlefish shell that has been lying in the terrarium or pen for a longer time will no longer attract the attention of the

Cuttlefish shell is an excellent source of calcium

Photograph Ricarda Schramm

Egyptian Tortoises whereas freshly fragmented pieces or grated powder are taken note of immediately and often consumed on the spot. Cuttlefish shell soaked in rainwater for a few days is given preference, too.

Algal lime: Some keepers have reported that the tortoises in their care would not touch cuttlefish shell. They had to resort to a very good alternative in the shape of powdered algal lime, which is, experience has shown, consumed with a great appetite. This source of calcium is particularly recommendable for use with juveniles whose beaks will not yet be strong and hard enough for biting off pieces from the rather solid cuttlefish shells at first. Algal lime is processed from the deposits of red algae and aside from calcium carbonate also contains various trace elements.

Algal lime is not to be sprinkled over other food either, but should be made freely available in low bowls. As described above for cuttlefish shell, the tortoises will lose interest in algal lime after the powder has been standing for some time. It is therefore advisable to replace the contents of the bowls with fresh powder once every week. It will then be an interesting observation how all tortoises of the terrarium rush to converge on the fresh powder and ingest it with gusto.

Algal lime is available from a range of commercial manufacturers.

Algal lime is another good source of calcium.

Photograph Ricarda Schramm

Fresh algal lime powder is very popular amongst Egyptian Tortoises.

Photograph Ricarda Schramm

The traces of a good intake of algal lime just cannot be overlooked.

Photograph Ricarda Schramm

Eggshells: Many keepers use the crushed shells of chicken eggs as a source of calcium for their tortoises. There are, however, indications that tortoises will absorb this type less effectively than the calcium from the other sources mentioned before.

In contrast to mammals including humans, tortoises do not access nutrients through the action of a strong acid in the stomach, but rather with the aid of enzymes from micro-organisms in the intestine. They have no cells in the stomach tissue that would produce hydrochloric acid, and the pH in their digestive tract is significantly higher than in mammals (JENNEMANN 2010) – the lower the pH, the more acidic a liquid will be.

The shells of chicken eggs are much thicker and harder than those of wild birds. They contain so-called defensines, which are antimicrobial proteins that serve the defence against bacteria, fungi and toxins. While some of these defensines are broken up when eggs are boiled, tortoises will find eggshells difficult to digest further due to their lack of stomach acid. This is also illustrated by the fact that shells of chicken eggs cast onto the compost heap can still be found there a few years later – composting is likewise based on the action of micro-organisms. It is therefore not surprising that particles or even fragments of eggshells can be found in the faeces of tortoises, and some veterinarians have come to warn against the possibilities of intestinal obstruction and injuries to the intestinal walls through the sharp-edged fragments of the shells (e.g., PRAUSER 2013). The latter risk could be avoided with some effort by pulverizing the eggshells in a grinder first. In general, enlarging a surface will also enlarge the attack surface for intestinal metabolic processes, improving the degree of exploitation. This is already facilitated by cuttlefish shell due to the inclusion of a myriad of tiny air bubbles (it serves as a buoyancy body in squid) and the pulverized form in the case of algal lime. Why should one therefore take the pains of first pulverizing eggshells if they do not have any advantage over the easily obtained cuttlefish shell or algal lime?

Vitamin Supplements

The terrarium supply trade offers a host of vitamin and mineral supplements for reptiles that are supposed to be sprinkled over their food. Many of these also contain calcium. These preparations cannot be considered a feasible source of calcium for tortoises, however, because to cover their needs of this mineral, you would have to administer irresponsibly large amounts of them, and this would entail a gross overdosing of the vitamins included. This risk of overdosing is exactly what many veterinarians specialising in reptiles warn against

(Dennert 2008, Kölle 2009, Kölle & Blahak 2015, Steidele & Friz 2014, Sternberg 2017). An overdosing of vitamin A, for example, can lead to large parts of the skin detaching, exposing reddened, oozing tissue. These wounds will then open the doors to various pathogens that cause secondary infections with the distinct potential of eventually killing the animal. One such example is the Central Asian Tortoises figured below, which unfortunately did not survive the consequences of vitamin overdosing. The mentioned authors all point out that it will be absolutely unnecessary to add any vitamin supplements if tortoises are kept according to their biological needs, fed a wide variety of wild herbs, and supplied with calcium and UV-B to an adequate extent.

Large patches of skin have detached as a result of an overdosing of vitamin A in a Testudo horsfieldii.
Photograph Ricarda Schramm

The same phenomenon: overdosing of vitamin A has caused large patches of skin to come off, leaving reddened and oozing tissues exposed in a Testudo kleinmanni.

Photograph Joe Stammet

It is only if behavioural anomalies such as apathy or other physical changes suggest there might be a health issue developing that a veterinarian specialising in reptiles should be consulted to find out what the problem is and left to design an adequate therapy with suitable medication or supplementation.

A *Testudo kleinmanni* experiencing a rainy day.

Photograph Franziska Biedenweg

Water

Rain showers: In their natural habitats, Egyptian Tortoises will experience fairly profuse rains only in the months from September through April, with precipitation events commencing haltingly in September and remaining fairly weak at first, increasing towards December, and decreasing once more from January through May at the latest. The highest amounts of precipitation are seen in the months of December and January, whereas virtually no rain will fall from June through August. Even heavy rain will not normally cause puddles to form, as the water tends to dissipate into the sandy ground almost immediately. Bodies of standing water are absent and do not offer opportunities to drink and cover the need for water. The proximity of the sea creates high levels of relative humidity, however, even in the months without rain, causing dew to precipitate at night. In contrast to some other species of *Testudo* Egyptian Tortoises are unable to cover their need for water by drinking from water bodies that will be available more or less all the time. *Testudo kleinmanni* had to find other means of adapting to its extreme environment by developing specialised strategies and behavioural patterns. It has learnt to take in water drops during and after

Raindrops are sucked up from plants.

Photograph Ricarda Schramm

Testudo kleinmanni will raise the rear portion of its shelled body in order to let rainwater run towards the mouth or extend the head towards the rain.

Photographs Franziska Biedenweg

rain showers or make the most of the morning dew – drops that accumulate on plants, rocks or even the shells of other tortoises.

Some tortoises will position themselves right into the direction of the rain and extend the head towards the raindrops with the mouth open. Others will raise the rear portion of the body on extended hind legs so that the rainwater will run down the shell towards the head where it is then taken in. This so-called "fog-basking" behaviour is also known from the head-stander beetle (*Onymacris unguicularis*) from the Namib desert, which will in this manner take in water droplets precipitating from fog.

Knowledge of this instinctive behaviour of *Testudo kleinmanni* explains the observations by some keepers that a water bowl will be made use of only rarely, if at all, and only by some individual tortoises (Branser 2011a, b, Connor 1993, Fritzsche & Fritzsche 2002, Schmidt 2009, Tabaka & Senneke 2003, 2004). Simply put, Egyptian Tortoises do not know "what to make of it". If you want to use a water bowl nevertheless, you will have to select one with a very low rim that can hold only a minimum of water, because instances of drowning in a water bowl in the absence of the keeper have also been reported (various verbal comms.). A creative alternative that facilitates the instincts of these tortoises exists in the shape of flat "rocks" with an irregular surface made from clay in whose depressions a little water will collect during misting.

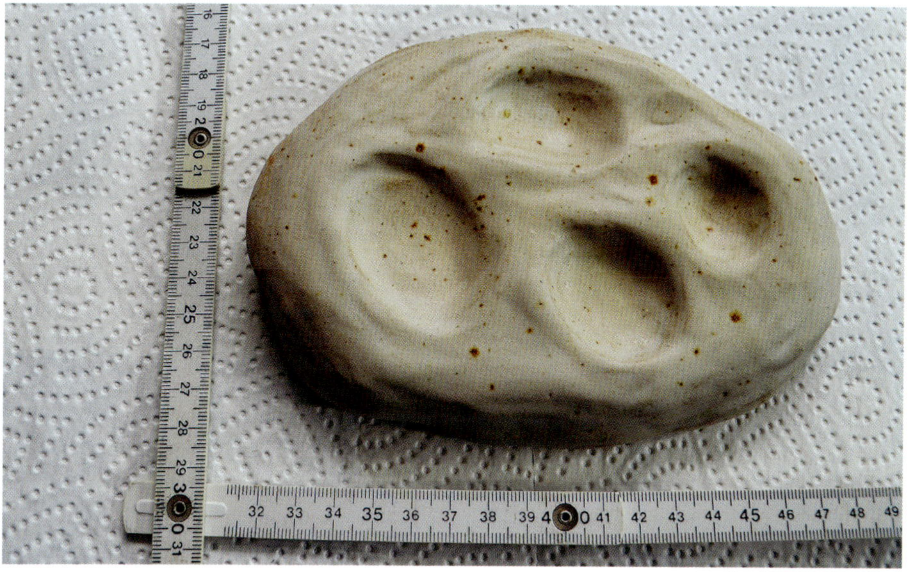

A pottered plate with depressions. Photograph Ricarda Schramm

A "rock" with low depressions that is used for drinking during emulated rain showers in the terrarium.

Photographs Ricarda Schramm

The same effect can be created by using individual, small, fired-clay dishes with diameter of at maximum 4–5 cm and an inner depth of not more than 1–1.5 cm that are sunk into the ground and so form “natural” depressions. Emulated rain showers motivate the Egyptian Tortoises to drink and they will then also drink from these artificial “puddles”. Plates or dishes serving this purpose are not available from commercial sources, but will have to be made to your specifications by a pottery.

Aside from creating opportunities for drinking, emulated rain showers also serve to moisten the bottom substrate and to create the required microclimate, enabling the tortoises to absorb water via the skin when they take to their underground shelters.

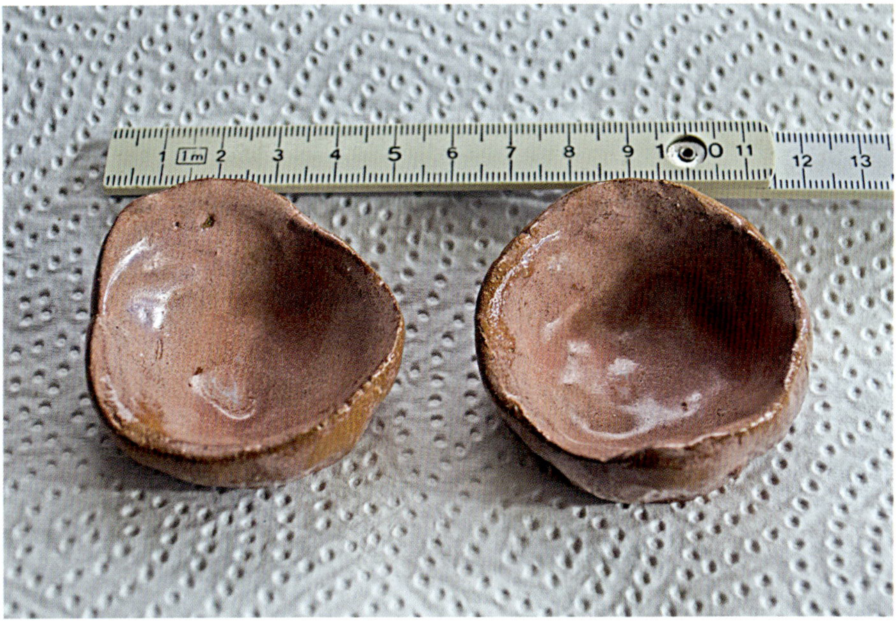

Small fired-clay dishes can be used to emulate puddles that will form during raining sessions and enable Testudo kleinmanni to drink.

Photograph Ricarda Schramm

Egyptian Tortoises use small fired-clay dishes to drink from during raining sessions.

Photographs Ricarda Schramm

Rain showers are most commonly created by means of a handheld flower mister or, preferably, a larger-volume pressure mister. They should in general be rather prolific on every occasion. An alternative may be the installation of a raining system of the type available from the specialised terrarium supply trade. The frequency of the emulated rain showers is adopted from the climate data existing from the home countries of the Egyptian Tortoises. Beginning in September with the first rains that signal the end of their aestivation period,

Left: A handheld flower mister with a capacity of 1.25 l;
right: A pressure mister with a volume of 5 l.

Photographs Ricarda Schramm

their frequency and extent are increased towards December, only to gradually decrease once more from January through May. In the wild, the beginning of May will typically see the last and little-prolific rainfalls before these eventually cease altogether during the further course of this month. No rain at all will then fall in the natural habitats of the Egyptian Tortoise from about mid-May through August which is when the animals aestivate. What will remain is a relatively high level of humidity, which needs to be emulated and maintained by means of foggers in human care (see p. 108).

Bathing: As a result of its adaptation to the semidesert climate of its home countries, *Testudo kleinmanni* has developed mechanisms for recycling a maximum of water from its intestinal tract and urinary bladder. As a consequence, it will dispose of only a minimum of urine. Because the tortoises will take in rather little water at the same time, there will be a risk of their not disposing of urates to an adequate extent and these urates forming bladder stones (GAD 2001, KÖLLE 2009, KÖLLE & BLAHAK 2015), which will then be difficult to excrete (see p. 225). It is for this reason of vital importance to ensure that they will take in enough water. With their small body volume, juveniles are particularly prone to dehydration and an eye therefore has to be kept on their drinking and absorbing moisture to a sufficient extent.

Emulated rain showers will also motivate Egyptian Tortoises to dispose of urine and urates.

Bathing the tortoises in lukewarm water has been repeatedly suggested as a support measure for taking in water and disposing of urates, which must but take place in a very low bowl (BRANSER 2011a, b GAD 2000, 2001, 2005, GRUBER 1993, 1998, REINARZ 2008, SCHEFFEL 2014, TABAKA & SENNEKE 2003). These bathing sessions should be facilitated once every week or fortnight throughout the activity season, commencing in September right after and ending in May right before the aestivation period. Each bathing session should then

Rain will motivate Testudo kleinmanni to dispose of urates.

Photograph Ricarda Schramm

Disposing of urates during a rain shower by a juvenile Testudo kleinmanni.

Photograph Ricarda Schramm

take some 15–20 minutes, possibly with replacing the water in between if the tortoises defecate or if the water cools down too much. Even if it then does not make the tortoises drink in the process, they will absorb water via the cloaca.

Other keepers will bathe their Egyptian Tortoises only right before and after the aestivation period or do entirely without it.

Bathing in lukewarm water

Photograph Franziska Biedenweg

Summer Dormancy (Aestivation)

The annual life cycle of the Egyptian Tortoise includes a period of aestivation, and any effort to keep *Testudo kleinmanni* in a naturalistic setting that strives to accommodate its biological requirements must therefore facilitate this period as well. Even though it may frighten newcomers in particular and leave them to wonder whether their precious little tortoises will survive this period of summer dormancy, courage and strong nerves are part and parcel of keeping this species. Adult *Testudo kleinmanni* depend on the sequence of alternating climatic conditions during the course of the year for synchronising the hormone levels that will see both males and females being ready to mate in spring. Aestivation is a key factor in this sequence, equally in importance to the period of winter dormancy (hibernation) of Mediterranean tortoises. While hibernation and aestivation are often equated, there are major differences between the two types of dormancy period. Winter dormancy sees all bodily functions reduced to a minimum, with heartbeat and breathing being slowed and the body temperature corresponding to the low values in the hibernation quarter. Without a rise in ambient temperatures, no active interruption of the dormancy and no intake of food or water will be possible in this scenario. Aestivation, on the other hand, will see metabolic processes being somewhat reduced, but the body temperature remains at a level at which the period could be interrupted and food and water be taken in. Triggers for the start of aestivation are rising temperatures, the ceasing of rains, and the disappearance of feeder plants as a result of both. The responsive process is then probably controlled by hormones. The month of May sees the hours of daylight and sunshine as well as temperatures increase while no more rain will fall. The highest values are eventually reached in July and August with a maximum of 14 hours of daylight, 12 hours of sunshine, and temperatures of 38 °C or more. As temperatures start decreasing and rains resuming around the middle or end of September, Egyptian Tortoises respond to the changing conditions with terminating their period of summer dormancy and becoming fully active once more.

Aestivating Egyptian Tortoises retain a small degree of activity in a captive setting. They may leave their shelters to move to another retreat or search for a little food and water in the cooler hours of morning and evening. Males and juveniles up to an age of about three years will remain more active than females at this time. It is therefore critical to make some dry feed and water in artificial "puddles" available at all times during the aestivation period.

Testudo kleinmanni ***aestivating in their shelters.***

Photographs: Top: Ricarda Schramm bottom: Franziska Biedenweg

The Course of the Year

The preceding chapters have detailed all key environmental parameters that need to be considered, emulated and controlled if *Testudo kleinmanni* is to be kept in a naturalistic manner that will fulfil all its biological needs. This information is here summarized for ease of reference.

Photoperiod: The hours of daylight equal the period of time during which basic illumination will need to be provided by means of lamps that are as bright as possible. This period is easily controlled via simple timers. The latter can be set to emulate dawn and dusk by switching the individual lamps on and off in a staggered manner, and dimmable lamps can be run via automated timed dimmers.

December has the shortest days with only 10 hours of daylight. The photoperiod should then rise gradually to reach 14 hours in July and August, only to decrease once more from September.

Hours of sunshine: Sunshine here equates to increased warmth and intensity of UV radiation and not to the hours of daylight provided through basic illumination. It is emulated by employing heat lamps with UV emissions with the aim of enabling the tortoises to bring their body temperature to "operating levels" and to expose themselves to vital UV radiation at the same time. These lamps are controlled with timers as well.

Like the hours of daylight, the periods of hours of sunshine are shortest in December and January at 6 hours, and longest at 12 hours in July, only to thereafter decrease once more towards December.

Temperature: The ambient temperature greatly depends on the placement of the terrarium and its type, i.e., whether it has a closed or open top, and must therefore be assessed individually. Very important here is to ensure that there will be a major difference between day and night temperatures. Reaching the required low temperatures of winter may prove difficult in certain situations, which suggests setting the terrarium up in a room that will not be heated during the winter months. If it can be placed only in a room that is heated during the day, the heating system should at least be turned off at night. Individual living situations may require that parts of the cover and/or front glass panes be removed or opened. Leaving doors connecting to unheated adjacent rooms or passages open or windows at a tilt may also help to cool the terrarium down. Draughts have to be strictly avoided, though!

The necessary day temperatures are created by means of heat and UV lamps – just as the sun would in the wild. Reaching them may require the installation of additional heat lamps without UV emission. Infrared lightless heat lamps, like Elstein ceramic heaters, may offer the required qualities as they can be used for controlling temperatures also at night. All heat lamps are managed via thermostat timers that will also facilitate different day and night temperatures. The very high values during the summer months are a key to initiating the aestivation period.

The lowest temperatures are to be provided in December and January, which is when values should not exceed 18–20 °C during the days and decrease to 10–15 °C at night. Temperatures then have to rise gradually and reach their maxima of 35–38 °C during the days and 25–26 °C at night in August before they are equally gradually reduced to said minimum in December.

Humidity: Egyptian Tortoises need a high level of relative humidity that will fluctuate little during the course of the year. This is achieved by means of foggers controlled electronically with timers that permit to provide different levels for day and night. The reservoir tanks of these devices are filled with distilled water to which may be a added a small amount of dissolved pure sea salt. This will create the salt-laden humid air that corresponds to the conditions near the sea in the natural habitats. Reaching and maintaining the required levels of humidity is again influenced strongly by the ambient conditions at the place of the terrarium. If the ambient air is generally dry, the foggers will have to work harder.

From October through April, the level of relative humidity should be set to 55–65 % during the days and 70–80 % at night. From May through September, the hygrometer should show values of 50–60 % during the days and 65–75 % at night.

Rain showers: Prolific rainfall events are emulated with the aid of a pressure mister from September through April. An electronically controlled raining system may be employed as well.

Beginning in September, the emulated rain showers are gradually intensified and increased in frequency to their maximum of two to three times per week in December and January, which are the months with the highest amounts of precipitation in the wild, and are then gradually reduced to once a month in April. Emulated rain events enable the Egyptian Tortoises to drink by taking in raindrops from plants, rocks and their own shells. Their intake of water can be promoted further by bathing them occasionally in lukewarm water of little depth. No rain showers at all are to be emulated from May through August in order to facilitate a period of summer dormancy.

Food: A naturalistic diet will consist mainly of wild herbs that are poor in proteins, but rich in fibre and calcium. The amount and composition of the food for Egyptian Tortoises varies with the seasons of the year, emulating the pattern in the natural habitats as well.

The month of October is when fresh herbs are first to be added to the dry feed, with the numbers of days with fresh feed and the amounts offered being gradually increased from week to week. Fresh herbs are then offered on a daily basis from November through April. Leftovers of wild herbs from the previous day need not necessarily be cleared away, but can be a left as a source of drying food if they are still unsoiled with faeces or urine. Adult *Testudo kleinmanni* will readily go without fresh feeds for a day in between, but juveniles should be given fresh herbs every day during their activity period.

The month of May will then see a gradual reduction in the number of days and amounts of fresh feed, following in reverse the feeding pattern of October, and no fresh feed will be offered from June through September.

Dried plant matter and flowers should be available to Egyptian Tortoises all year round, even when they actually aestivate, because they may occasionally leave their shelters to consume some.

Annual care schedule: We have compiled all the preceding parameters per month in a tabulated overview for ease of reference on the following three pages.

What you should know: The Annual Care Schedule can be found as a separate insert in the back of the book so that you will have it handy whenever you may need to consult it.

Tables on the following three pages:
The Annual Care Schedule for Testudo kleinmanni

Month / Parameters	January	February	March	April
Hours of daylight	10–10.5 h, increasing over the month	10.5–11 h, increasing over the month	11–11.5 h, increasing over the month	11.5–12 h, increasing over the month
Hours of sunshine, heat lamps	6 h	7.5 h	7.5 h	8.5 h
Temperature day / night	18–20 °C 10–15 °C	20 °C 10–15 °C	23 °C 15 °C	24–26 °C 20 °C
rel. humidity day / night	55–65 % 70–80 %	55–65 % 70–80 %	55–65 % 70–80 %	55–65 % 70–80 %
Rain showers	2–3 per week	1 per week	1–2 per month	1 per month
Feed	fresh feed daily, dry feed freely available	fresh feed daily, dry feed freely available	fresh feed daily, dry feed freely available	fresh feed daily, dry feed freely available

Month / Parameters	May	June	July	August
Hours of daylight	12 h	13 h	14 h	14 h
Hours of sunshine, heat lamps	9.5 h	10 h	12 h	11.5 h
Temperat-ure day / night	26–30 °C increasing over the month 22 °C	30–33 °C increasing over the month 22–24 °C	33–35 °C increasing over the month 24 °C	35–38 °C increasing over the month 25–26 °C
rel. humidity day / night	50–60 % 65–75 %	50–60 % 65–75 %	50–60 % 65–75 %	50–60 % 65–75 %
Rain showers	0	0	0	0
Feed	Per week: fresh 1st: a little daily 2nd: 3 x 3rd: 2 x 4th: 1 x	no fresh feed, dry feed freely available	no fresh feed, dry feed freely available	no fresh feed, dry feed freely available

Month / Parameters	September	October	November	December
Hours of daylight	13–12 h decreasing over the month	12–11.5 h decreasing over the month	11–10.5 h decreasing over the month	10 h
Hours of sunshine, heat lamps	9 h	8 h	7 h	6 h
Temperat-ure day / night	35–32 °C decreasing over the month 24 °C	30–26 °C 25–22 °C decreasing over the month	24–26 °C 17–20 °C	18–20 °C 10–15 °C
rel. humidity day / night	50–60 % 65–75 %	55–65 % 70–80 %	55–65 % 70–80 %	55–65 % 70–80 %
Rain showers	first shower in the last week	2–3 per month	1–2 per week	2–3 per week
Feed	no fresh feed, dry feed freely available	Per week: fresh 1st: 1 x 2nd: 2 x 3rd: 3 x 4th: a little daily	fresh feed daily, dry feed freely available	fresh feed daily, dry feed freely available

Propagation

The Giza Zoo in Cairo, Egypt, was the first to propagate *Testudo kleinmanni* in larger numbers in the early 20th century (Buskirk 1985, Flower 1933). in larger numbers in the early 20th century (Buskirk 1985). First successful breeding attempts in Europe were eventually registered by the Jersey Wildlife Preservation Trust (today, Jersey Wildlife Conservation Trust, with its head office in the Jersey Zoo on Jersey Island, GB) in 1990 (Gibson 1997), and the hatchlings produced by the Tortoise Trust in November of 1994 were likely the first on the British mainland (Highfield & Martin 1995, 2001). Manfred Rogner (1996) reported on further successes. In the US, Lee Hobbs is considered the first keeper of Egyptian Tortoises who managed to also propagate them in 1992 (Hobbs & May 1993, Hoekstra 1994). In Germany Jürgen Gad (2000, 2005) und Hans-Jürgen Gruber (1998) were amongst the first to successfully breed this species in the late 1990'. Thanks to improved technological means becoming available and much increased specialist knowledge, its successful captive propagation eventually became more widespread only in the 21st century, both amongst private enthusiasts and organisations overseeing breeding programmes on a global scale (e.g., the European Studbook Foundation). *Testudo kleinmanni* has nevertheless remained a rarity and the comparatively few captivity-produced young cannot as yet cover the demand for this small tortoise, which is evident from the fact that illegal exports and attempted imports have not stopped to this day.

Preconditions:

Like with other species, the successful captive propagation of *Testudo kleinmanni* requires first and foremost that potential breeder specimen be kept in optimal conditions (Vinke & Vinke 2004, 2006).

This includes the following parameters:

- A harmonious colony of tortoises representing both sexes must be present.
- Husbandry conditions must be optimal, as detailed in the preceding chapters.
- The environment must be stress-free.
- Suitable oviposition spots must be available.
- An incubator must be available.

Assembling a breeding group: It usually all begins with the acquisition of some juveniles. Like with all species of *Testudo*, the sexual identities of these juveniles will remain a puzzle for some time until they have grown to a certain age or size and can eventually be sexed. As is the case with many reptiles, the sex of an embryo is determined by the temperatures under which it is incubated.

In the present case, higher incubation temperatures will produce more females and lower ones more males. Some breeders will therefore incubate their eggs at certain temperatures with the aim of producing more of the one or other sex. While this method is not a hundred percent failsafe, it will at least create a greater likelihood. A colony of juveniles that grows up together is very likely to grow into a harmonious colony of adults, because its individuals will know each other from baby age. This cannot necessarily be supposed if a colony is assembled from tortoises that are adult already.

Another important factor is the sex ratio in a colony (see p. 31 ff for sex-indicative traits). The number of males should always be lower than that of the females. Successful breeding is often based on a sex ratio of one male and two females (breeders commonly refer to such a constellation as "1,2" with the first digit specifying the number of males and the second that of the females in a breeding group; (BRANSER 2011a, b), 1,3 (HIGHFIELD & MARTIN 1995, 2001, REINARZ 2008) or 1,4 (SCHEFFEL 2014), but there have also been reports on scenarios where keeping a pair (1,1) produced the desired results (BRANSER 2011a, b, SCHEFFEL 2013, SCHMIDT 2009). The great risk in keeping the same numbers of males and females, or even fewer females than males, is that the female(s) will be unable to escape the incessant attentions of the males and thus be stressed.

In contrast to many other species of *Testudo*, male *Testudo kleinmanni* will behave relatively gently amongst each other, even when rival males engage in combats over the favours of a female (HOBBS & MAY 1993). This renders it possible to have two males living in a colony, provided there will be a correspondingly large number of females and that the colony is accommodated in a large and well-structured enclosure.

Reaching sexual maturity depends more on having grown to a certain weight than to a certain age in Egyptian Tortoises, as will be the case in other species of *Testudo*, too. There are at present no realisations at which weight and age females will first start mating and depositing eggs in the wild. Field studies suggested on the basis of growth rings that ovipositing females were between six and nine years of age, but it remained unknown whether these females were "first-timers" or not – they were very likely not (ATTUM et al. 2011, GEFFEN & MENDELSSOHN 1991, WILSON et al. 2003; see also p. 63). What may be quite safe to suppose from observations in other species is that sexual maturity is reached earlier in human care than in the wild, simply because the animals kept in captivity will benefit from an optimised food supply and usually gain weight more quickly. Males kept in a captive environment will typically start showing an interest in mating on reaching a weight of about 100 g and an age of two years, in exceptional instances even a little earlier.

For their part, females in human care can be expected to produce their first eggs on reaching a weight of some 200 g at an age of three to four years (Biedenweg eigene Beobachtungen, Branser 2011a, b, Fritzsche 2002, Gad 2000, 2005, Hobbs & May 1993, Scheffel 2014).

Mating:

So far, very few reports exist on mating events observed in the natural habitats in late winter and early spring (Geffen & Mendelssohn 1991, Gruber 1998), with the mentioned authors supposing that these would be exclusive to this timeframe in the wild. The situation is entirely different in human care, though, where copulation events have been witnessed from October and November through May, partly with breaks in between (Branser 2011, Flower 1933, Gad 2000, 2005, Highfield & Martin 1995, 2001, Hobbs & May 1993, Mendelssohn 1982, Reinarz 2008, Scheffel 2014, Stettner 2004). Own observations in captivity recorded mating events from mid-October through the end of February, males trying to copulate for the first time at a weight of 80 g and an age of 1.5 years, and being successful (fertile eggs) at 100 g and 2.5 years (Biedenweg own obs.).

Like with all species of *Testudo*, a male of *Testudo kleinmanni* will leave no doubt of his intentions to mate with his chosen female. He will begin by pursuing the female and continually ram her shell, effect bites to her front and hind legs, and energetically circle around her in an attempt to make her stop in one place so that he can mount her from behind. The female will comply and stop running away to facilitate copulation only if she is conceptive. This means that it will typically take several such attempts by the male before mating can occur. Once the female signals her consent by stopping in one place will the male be able to locate the female's cloaca with his tail and insert his penis. The female will then even actively facilitate copulation by lifting the rear portion of her body and remaining still. Sometimes, it may even be observed that the female will retract her head into the shell for a few seconds, making her cloaca protrude a little farther from the rear opening of the shell and rendering it more readily accessible to the male. The penis of a male has a remarkable length relative to his small body dimensions and may measure about two thirds to almost the total length of the plastron with a thickness of some 3 cm erreichen (Biedenweg own obs.; Scheffel 2014). A mating event may take up to 30 minutes (Scheffel 2014, Highfield & Martin 2001). The characteristic sounds produced by male *Testudo kleinmanni* during copulation are entirely different from those emitted by any other species of *Testudo*. Difficult to describe in words, the males will produce high-pitched, cooing and chirping noises that one needs to have heard to believe that they can be produced by a tortoise.

Not every copulation event will be successful and lead to fertile eggs being

The male will first court a female before copulation can ensue in Testudo kleinmanni.

Photographs Franziska Biedenweg

Top: Males will sometimes extrude the penis when bathing.
Bottom: The penis of a male
***Testudo kleinmanni* can be remarkably large.**

Photographs Franziska Biedenweg

Mating in Testudo kleinmanni.

Photograph Franziska Biedenweg

produced. Assessing the success is furthermore hampered by the fact that females are able to store functional sperm over an extended period of time, as has been demonstrated by clutches of fertile eggs that were laid two to three years after a male had been separated from a female, which is a phenomenon also known from many other species of *Testudo*. Separating the sexes for some time or increasing the level of competition by keeping two males in a breeding colony may increase the intensity of mating activities and thus the breeding success (GAD 2000, 2005, SCHEFFEL 2014).

Oviposition:

Hardly any information exists as yet on the timing of oviposition in the natural habitats either. GEFFEN & MENDELSSOHN (1991) reported on four females that were x-rayed and found to carry shelled eggs between March and June. Clutches were found in early July. GRUBER (1998) reported on one to three clutches that were deposited at intervals of 20–30 days, usually in April and May, but sometimes already in late March and up to early June.

Like in the case of mating events, observations on the timing of oviposition events in captive settings differ significantly from those in the wild, with examples ranging from May through July (Hobbs & May 1993), largely from March through May (Gad 2000), between November and May (Fritzsche & Fritzsche 2002), from about January (Reinarz 2008), from February through May (Stettner 2004), to mid-February through mid-June (Biedenweg own obs.), usually some 4–6 weeks after mating took place.

If you know the tortoises in your care and their behaviour intimately, you cannot help but notice the changes in the appearance and behaviour of gravid females. Burdened with producing eggs, they will feed much more than they normally would. They will also exhibit a greater need for calcium by taking in large amounts of cuttlefish shell, algal lime etc. in order to be able to shell their eggs properly. The increased intake of food plus the growing eggs will see them gain weight rapidly and distinctly. Some time prior to eventually depositing their eggs you may notice that they will keep the xiphiplastron in a flipped-down position, which creates more space for the eggs in the body cavity (see pp. 16, 28).

The xiphiplastron is clearly flipped down in a female *Testudo kleinmanni* shortly before oviposition.

Photograph Franziska Biedenweg

For comparison:
Left: the position of the xiphiplastron in a female not carrying eggs in her body cavity;
right: as the xiphiplastron is kept flipped down by a female shortly before depositing eggs.

Photograph Franziska Biedenweg

As the point of depositing her eggs draws close, the female will eventually start taking in much less food. She will also become more and more restless the closer the point of oviposition comes. She may then be found cruising the environment in the first light of early morning, excavating test nest pits in various spots throughout the terrarium. Initially digging holes only with her front legs, she will examine the consistency of the soil, temperatures and moisture levels. The female will also often be seen checking out the soil by extensively sniffing it. Selecting a spot that promises optimal incubation conditions will sometimes take her a long time, because it is basically the only thing a female can do for her offspring in the absence of any further maternal care.

During the period right before oviposition, gravid females may present themselves quite incompatible with their conspecifics and fend off courting males with emphasis. Even other females are kept at bay through dominant behavioural expressions such as biting and mounting (Scheffel 2014,

Vinke & Vinke 2004). A highly gravid female may sometimes remain in the oviposition spot she has selected and even spend the night there buried headover into the sand, probably in order to keep other females away from it (Biedenweg own obs.; Scheffel 2014).

The few existing reports on oviposition events in the natural habitats suggest that female Egyptian Tortoises have a predilection for shaded areas facing south or east under bushes (see p. 63; Geffen & Mendelssohn 1991, Gruber 1998). This is probably where moisture levels in the ground are sufficiently high and temperatures will not reach dangerously high values due to the shade cast by the bush. This observation matches those made in captivity where females likewise tend to select oviposition spots at the bases of plants, below a piece of root, in the entrance area of an underground shelter, or in similar situations that offer some cover and usually lie within the peripheral range of a heat lamp (Biedenweg own obs.; Hebbeler pers. comm., Reinarz 2008, Schmidt 2009, Scheffel 2014). It is thus important for the keeper to ensure that these potential oviposition spots are kept adequately moist and so enable the females to excavate proper nest pits.

Two female Testudo kleinmanni busy depositing eggs within the range of a heat lamp, one at the base of a plant, and the other in the entrance area of a retreat.

Photograph Michael Hebbeler

When environmental conditions are deemed right and the time has come, the female will eventually start excavating the nest pit with her hind legs.

A female starts excavating her nest pit using her hind legs.

Photograph Franziska Biedenweg

GEFFEN & MENDELSSOHN (1991) reported on nest pits of 3–4 cm deep in the wild. In human care, these pits may well reach up to 5 cm deep (GRUBER 1998). The low depth of the pit is of course a result of the small body size of a female Egyptian Tortoise and will be as deep as a female can reach with her hind legs. She will again and again feel out the dimensions of the pit with her hind feet, assuring herself eventually that it is large enough for her clutch. Oviposition then commences with the extrusion of the first egg. Labour is actively supported by the female retracting her head as deeply into the shell as it will go and sometimes audibly pressing all air from her lungs. It may take several such pressing contractions for the egg to exit from the birthing canal, in particular in the case of the first egg.

Each egg laid in this manner is gently and carefully positioned in the nest pit and then lightly pressed in place before the next egg will be laid. Once all eggs of a clutch have been deposited thus, the female will wipe the excavated sand back into the nest pit with her hind legs until it is closed and then additionally compact the soil through stamping. In spite of the strenuous efforts required, this is done

The first egg exits the cloaca.

Each egg is gently placed in the right spot with the hind legs.

Photographs Franziska Biedenweg

The nest pit is closed carefully with sand.

Photograph Franziska Biedenweg

with extreme care and attention to detail so that nothing will afterwards point to the existence of a clutch in this spot (Biedenweg own obs.; Scheffel 2014).

The eggs are surprisingly large compared to the small body sizes of females. Measuring 25–28 mm, more rarely as much as 30 mm, in length, having a width of 20–22 mm, more rarely up to 23 mm, and counting a maximum of three eggs per clutch, they can equal up to 28 % of the carapace length of a female (Gruber 1998). A comparable size ratio is found only in the eggs of the Central Asian Tortoise, *Testudo (Agrionemys) horsfieldii*, which likewise lives in arid (dry) regions with extreme environmental conditions (Fass 2013). This fact suggests that eggs that are supposed to successfully incubate in these circumstances need to be furnished with as much in nutrients in the shape of yolk as is possible in order to maximise their chances of survival (Hartmann 2009).

Most breeders have counted one to three clutches that are deposited at intervals of 20–30 days and comprise 1–3 eggs each per female and year (Biedenweg own obs.; Gad 2000, Hobbs & May 1993, Schmidt 2009, Stettner 2004) but four eggs in a clutch have also been noted in some exceptional instances (Highfield & Martin 1995, 2001), as have up to four clutches of up to four

Right:

Once the nest pit has been closed, nothing betrays the fact that eggs have been deposited here.

Bottom:

Carefully retrieving the clutch.

Photographs Franziska Biedenweg

eggs (Biedenweg own obs.; Gad 2005) and even more (Branser 2011a, b, Scheffel 2014).

After the female has completed her task and moved off, the eggs need to be retrieved with great care and transferred to an incubator. Here it is of vital importance to not turn the eggs in the process, but to keep them in the exact same position in which they were laid and place them thus into the incubation substrate. This is most easily controlled by marking their top spots each with a cross made with a soft pencil, which will also allow you to keep them thus when handling them later. The reason for this detail being so critical is that the yolk in the egg will sink to the bottom of the egg and the embryo come to lie and grow in place on top of it within hours after the eggs having been laid (Ewert 1985). If the egg were turned over after this process is completed, the embryo would slip under the yolk where it will not normally be able to develop and perish as a consequence. Furthermore, a shift in position may cause the amnion between the embryo and the yolk membrane to be ripped off (Köhler 2004). This risk is greatest during the first third of the incubation period.

Marking the top spots of each egg prior to their transfer to an incubator.

Photograph Franziska Biedenweg

Some authors recommend for hygienic reasons to carefully wash the eggs before placing them in an incubator (Köhler 20014, Vinke & Vinke 2004). However, it has since emerged that the thin slimy film covering them has an antibacterial action that would be removed when washing them (Jasser-Häger & Winter 2007, Knappe 2011). It is therefore entirely sufficient to brush off adhering sand using the fingers or a soft cloth. As a matter of principle, hands should be hygienically clean whenever the eggs are handled, and using gloves may be a good idea.

Incubation of Eggs:

There is once more only one report providing information on temperatures and humidity conditions in the nest pits in the natural setting (Geffen & Mendelssohn 1991). These authors measured the temperature in the sand at a depth of 5 cm, revealing it to be 24.3 to 38.2 °C, in July, 23.5 to 35.3 °C in August, and 23.9 to 34.5 °C in September. Moisture levels of the sand at this depth were also analysed, but only as the difference in weight prior to and after drying, which does not unfortunately provide pointers that could be adopted for the required humidity values in an incubator. This difference amounted to between 0.16 and 0.27 % and actually demonstrates that there is very little water contained in the sand (see p. 63). Artificial incubation should therefore be conducted on a substrate that is relatively dry rather than too moist. Humidity values that are too high, overly frequent remoistening the substrate, or even water precipitating on the lid of the incubator and dripping onto the eggs can cause the eggshells to crack, embryos to perish, or make hatchlings emerge prematurely and with exceedingly large remnant yolk sacs (Jasser-Häger & Winter 2007, Vinke & Vinke 2004). In their review, Irmtraut Jasser-Häger & Annemarie Winter (2007) noted that the eggs of the Central Asian Tortoise, *Testudo (Agrionemys) horsfieldii*, which lives in comparably arid, extreme habitats and spends the summer aestivating, were particularly sensitive to the impacts of moisture. Ude Fass (2013)) therefore incubated the eggs of this species in a dry substrate at a relative humidity of only 40–50 %. Considering the similarities in their ecology, similar incubation parameters may be supposed to apply to the eggs of *Testudo kleinmanni* as well. Comparisons with other tortoise species from marginal zones of deserts, e.g., with various species of gopher tortoises (*Gopherus* spp.), could possibly provide further insights, but the present book is not the place to undertake them.

Incubators: In the early days of propagating tortoises in captivity, no commercially manufactured incubators were available, necessitating that eggs obtained be incubated in self-made devices – and often quite successfully so (Schramm own obs.). The most widely applied technique at this time became known as the "aquarium method" (Budde 1980, Köhler 2004, Vinke & Vinke 2004) and it is one that has been practiced with success until today (Biedenweg own obs.). Its greatest drawback is, however, that it is notoriously difficult to control humidity levels, which may be way too high for some eggs, owing to the copious evaporation of water driven by a submerged aquarium heater (Vinke & Vinke 2004).

Today, however, the terrarium supply trade holds a multitude of tray, motorised and cabinet-type incubators that will all produce good to very good results. As a single example may here serve the very widely employed tray incubator FB50 manufactured by the German-based company Jäger und Pfrommer Brutapparate (aka. Jaeger Bruttechnik) that is run with a damp-incubation set. It has been used for many years to very successfully incubate the eggs of Mediterranean tortoises (Schramm own obs.), including those of *Testudo kleinmanni* (see Tables pp. 208-210). Whatever your choice of incubator it may be in the end,

Left: A tray-type incubator FB50 manufactured by Jäger und Pfrommer Brutapparate (aka. Jaeger Bruttechnik).
Right: Accessory damp-incubation set
Photographs Ricarda Schramm

it should be set up and running at the desired temperature and humidity levels at the latest when the first signs of a female readying herself for depositing her eggs, because it will usually take a few days of running it in before all values have stabilised.

What you should know: The incubator should be placed in a dark room where it will be safe from shocks and vibrations.
Both the ambient temperature and relative humidity values in this room should be fairly unvarying and never exceed those set in the incubator.
More substantial changes in the room temperature and humidity may influence the values inside the incubator.
The chosen room must furthermore be kept free of potential harmful influences such as cigarette smoke, for example.

Incubation boxes: Principally speaking, all sorts of vessels may be employed as incubation boxes, preferably those made of plastic. A widely practised solution to this effect is cricket containers that will already be furnished with small ventilation holes in their sides. Cricket boxes usually come in two sizes, square at 11.5 x 11.5 x 6 cm or rectangular at 18 x 13.5 x 6 cm. Larger additional ventilation holes can be punctured along the top edge with punch pliers. It is recommendable to cover each of these incubation boxes with a small pane of appropriately sized Plexiglas in such an offset position that the four corners will remain open for ventilation. This will for one prevent that precipitating water can drip onto the eggs and, for the other, that an unnoticed hatchling may overclimb the edge of the container and fall from it.
It is highly recommendable in general to use a dedicated (and properly labelled) incubation box for every clutch, as this will prevent hatchlings emerging from older clutches from disturbing those eggs that were added later and possibly in this manner compromise their development.

An ordinary cricket container with additional, larger ventilation holes and a small cover pane of Plexiglas

Photograph Ricarda Schramm

Incubation substrates: Vermiculite is the incubation substrate of choice of many breeders. It is a silicaceous mineral with a lamellar structure that will be able to absorb and store huge amounts of water and subsequently release it slowly to the surrounding air. This means that the moisture content in the substrate will be about equal to the humidity level inside the incubator. It also means on the other hand, that if the humidity level in the incubator is too high, the substrate may also become rather too moist for the incubation of eggs of *Testudo kleinmanni*.

Taking once more a cue from nature, sand or a sand/loam mix lend themselves as the incubation substrates of choice (Vinke & Vinke 2004), as these will facilitate incubation in drier conditions; they are in fact often the key to successful incubation of *Testudo kleinmanni* eggs (see Tables pp. 208-210).

Breeders differ in opinion as to how the eggs are best placed on/in the substrate, ranging from lying openly on top of it, via half-buried, to fully buried in it. Openly incubated eggs bear the risk of a hatchling rolling its own egg, or neighbouring eggs, during its hatching process, causing it to come lying under the gelatinous contents of the eggs and drowning (Vinke & Vinke 2004). There will furthermore be no substrate that could have a buffering effect on temporarily excessive moisture levels. In the natural setting, the eggs will be securely held in place and provided with counter-pressure by the surrounding soil, and hatchlings need to dig themselves out of the ground first, during which process they will probably use up what is left of their yolk reserves. Eggs buried half in the substrate, on the other hand, may experience a difference in ambient moisture between the open and the buried halves, which may cause tension cracks to appear in the eggshell. Again, comparison with the situation in nature clearly suggests that fully buried eggs may be the most naturalistic approach. The drawback here is that checking on the eggs will hardly be possible, which is probably the reason why so many breeders opt for incubating their eggs out in the open instead, especially as it will lead to good results in the end as well. At the present stage, there still exists too little information to state with any degree of certainty which may be the best strategy for incubating the eggs of the Egyptian Tortoise, though.

Temperature and humidity: The sex of a tortoise is not genetically fixed right from the start, but will rather be determined by the incubation temperature during the embryonic development approximately towards the end of the first third of the incubation period. Incubation temperatures therefore play a major role when tortoise eggs are incubated artificially. As far as the eggs of the Egyptian Tortoise are concerned, we are still at a "stage of experimentation" with regard to temperatures and moisture levels. Still unknown in *Testudo kleinmanni* is the vertex temperature, which specifies the value at which males and females will hatch in equal numbers. Explained in a simplified manner, eggs incubated above the vertex temperature will produce more females whereas those incubated below it will see more males hatching. Vertex temperatures are known to increase with increasing latitude at which habitats are situated (EWERT et al. 1994). That of *Testudo hermanni* lies at 32.5 °C with certain subspecies and local forms from more northern locales having their vertex temperature at 31.5 °C (WEGEHAUPT 2006). Ude FASS (2013) found out that this value lies at 32.5 °C in *Testudo (Agrionemys) horsfieldii*. Because it will always take a few years before sexual identities can be discerned, experiments to this effect are notoriously lengthy. In *Testudo kleinmanni* Sabine BRANSER (2011a) reported that she obtained exclusively males from eggs incubated at 28.8–29 °C and only females at 32 °C. This suggests that when incubating the eggs of this species at constant values, setting the incubator to 32–33 °C would make sense if predominantly females are to be produced. Even higher values are known to increase the likelihood for anomalies, malformations and the loss of embryos. No embryonic development will be possible below and above certain threshold temperatures.

Owing to the facts that the sexual identity will be determined near the end of the first third or during the second quarter of the incubation period, but the horn scutes are formed already during the first three weeks (EWERT 1985, KÖHLER 2004, WEGEHAUPT 2006), it may be sensible to incubate the eggs at slightly reduced temperatures during the first and last thirds and at a higher value during the middle third of the incubation period in order to forestall scutellation anomalies, but still produce predominantly female hatchlings (GRUBER 1999 cit. fide STETTNER 2004, HIGHFIELD & MARTIN 1995, 2001, STETTNER 2004). If you have only one incubator, this strategy will be feasible only for clutches that are all at the same stage of development, otherwise you will need several incubators.

Incubation temperatures that fluctuate between higher values during the day and lower ones at night are a method favoured by some breeders and which comes close to the conditions eggs are exposed to in nature (FRITZSCHE & FRITZSCHE 2002, GIBSON 1997). In this case, the average temperature around which the values oscillate will determine the outcome. It may however be so low in some instances that the proportion of male hatchlings will increase (WEGEHAUPT 2006).

All the aforesaid clearly indicates that monitoring the relevant parameters is of utmost importance and this means high-quality precision measuring tools. Temperatures in the incubator can be measured with analogue or digital devices with sensors and should be taken right in the incubation substrate next to the eggs. Highly accurate values of the relative humidity are provided by a hair hygrometer. The latter needs to be kept calibrated, which is fairly simply achieved by wrapping the device in a wet towel for about an hour, then taking it out and adjusting as quickly as possible the calibration screw in the rear so that the gauge in the front will read 100 %.

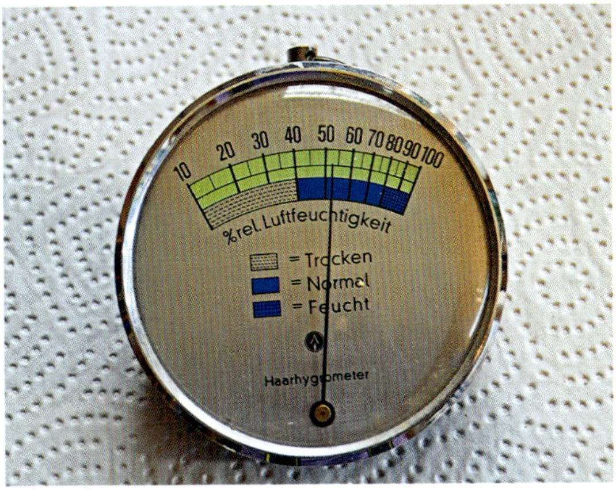

A hair hygrometer for measuring humidity values in the incubator

Photograph Ricarda Schramm

Realisations from the wild and from artificial incubation suggest that relative humidity values between 50 and 70 % will be practicable. There is no universal recipe with absolute values for temperature and humidity that would apply to the incubation of eggs of the Egyptian Tortoise. Too small are the numbers of successfully incubated clutches, and data cannot necessarily be directly compared if they were recorded under different incubation conditions. Whether captive propagation will be successful or not also always depends on how well the adult tortoises are cared for (Gad 2000). Overlooked clutches have repeatedly hatched in terraria (Gad 2005, Hebbeler pers. comm., Schmidt 2009).

An overview of incubation strategies applied by various authors for eggs of *Testudo kleinmanni* kept in human care is provided by the tables on the following pages. Comparing these data is hardly possible, however, because individual parameters may vary significantly between breeders and some informative data may not have been mentioned.

Tables on the following three pages:
Compilation of Literature Data on Incubation Strategies Applied for Eggs of Testudo kleinmanni

Author	Incubator	Substrate/ pos. of eggs	Tem-perature/ humidity	Incuba-tion period in days	Hatching rate/ hatchlings
Giza Zoo Kairo FLOWER 1933 BUSKIRK 1985	probably outdoors	probably sand/loam/ buried			69
Jersey Zoo, JWCT 1990 GIBSON 1997		sand	30 °C ((m) fluctuating 24 – 38 °C	85 – 95, – 110	5 males
HOBBS & MAY 1993	ordinary chicken egg incubator Hovabator	dry sand/ open on substrate	87 °F (30.5 °C)	96 – 123	50 %
HIGHFIELD & MARTIN 1995, 2001	"standard static air incubator"	vermiculite sandy soil or mix/ open on substrate	1st third 33 °C 2nd third 35 °C 3rd third 33 °C/ 70 – 80 %	78 – 111	
GRUBER 1. 1998 2.1999 Ang. nach STETTNER 2004	Jaeger	vermiculite	1998: 28 – 29 °C 1999: 28 °C/ 35 °C/ 28 °C per third	119 – 143 90	100 %/3 1

Author	Incubator	Substrate/ pos. of eggs	Tem-perature/ humidity	Incuba-tion period in days	Hatching rate/ hatchlings
Gad 2000, 2005	Jaeger	vermiculite half-buried	30.5 – 33 °C 70 – 90 %	78 – 98	35 % 43 % 1998 91 %
Fritzsche & Fritzsche 2002	in base-ment	sand, vermiculite open on substrate	fluctuating 27 – 34 °C 60 – 65 %	81 – 106	29 out of 31, 1 de-stroyed 1 infertile
Stettner 2004	Jaeger	vermiculite	1st third 28 °C 2nd third 33 °C 3rd third 28 °C	90 – 110	6
Wittmund 2005	Jaeger	vermiculite buried for two thirds	32 °C 70 – 80 %	1 egg opened after 121 days, live hatchling with yolk sac	1 out of 2, 1 de-stroyed
Reinarz 2008	drying cabinet Heraeus	sand, Seramis®, vermiculite open on substrate	31 – 32.5 °C 60 – 70 %	89 – 134	15 out of 20 2 destr. 2 infert. 1 perished

Author	Incubator	Substrate/ pos. of eggs	Tem-perature/ humidity	Incuba-tion period in days	Hatching rate/ hatchlings
Schmidt 2009	1. Jaeger 2. on-site in terrari-um	sand-loam 2. fully buried			1. 0 out of 8 2. 100 % 2 out of 2
Branser 2011a, b	1. indoor hotbed with hea-ting pad 2. Bruja	1. Sera-mis® 2. vermicu-lite buried for one third	28.8–29 °C (m) 32 °C (w) 50 – 60 % from 2011 70 – 75 %	85 – 130 (m) 87 – 142 (w)	2008 thru 2010 45 out of 114 51 infert. 17 per. 1 dest.
Scheffel 2014	Jaeger, va-rious other manufac-turers	vermiculite mostly open on substrate	Different tempera-tures for m and f	80 – 113 Ø 92 days all Ø 85 days (m)	
Dwor-schak homepage	Jaeger	sand-loam/ fully buried	fluctuating, cooler at night Ø 31.5 °C	70 – 90	
Biedenweg (own obs.)	1. Aquar-ium me-thod 2. Jaeger	sand-loam/ fully buried	32.5 – 33 °C 1. 90 % 2. 75 %	71 – 92	1. 3 out of 4, 1 infert. 2. 15 out of 17 fert., 1 infert., 1 per.

Embryonic Development:

Embryonic development scarcely differs with the individual tortoise species (Ewert 1985, Köhler 2004, Vinke & Vinke 2004, Wegehaupt 2006). Freshly deposited eggs are opaque to translucent and of a pale pink colour. Candling will now show the egg to be equally light throughout. A few days later, the yolk will have settled on the bottom while the egg-white floats on its surface, enveloping what is an early embryo. That the embryo has started to develop will be visible on the outside of the egg already at this stage: at first, a lighter or even white spot will appear that continues to spread around the girth of the egg, forming a band that is commonly referred to as a "belt" and then widens towards the poles until the entire egg will be white after about three weeks, but the exact speed at which this development takes place may vary.

The embryo will measure a few millimetres after about two weeks, about 1 cm after three, and fill half to three quarters of the egg already after four to five weeks. A very good and photographically illustrated overview of the embryonic development across various stages was provided by Gunda Meyer de Rojas (2018).

Left: A white spot has appeared on the egg six days after it was laid.

Right: The germinal disk (blastodisk) (top left with a red dot) on the yolk of an opened egg that has failed to develop further.

Photographs Franziska Biedenweg

From left to right:

• Right after oviposition • "Belting" after two days

• 22 days after oviposition – the egg has turned white throughout

Photograph Franziska Biedenweg

The development of the egg can be monitored nicely by shining a candling lamp through it. From about halfway through the incubation period, the embryo starts filling more and more of the egg's inner space until candling will no longer allow any light to penetrate. Eventually, an air bubble may form shortly before hatching, which indicates the spot where the amnion has come loose and which is where the hatchling will eventually crack the eggshell open for hatching.

What you should know: Frequent checks, handling and candling of the eggs may endanger the development of the embryos and in the worst-case scenario cause them to perish. This risk is greatest during the first two thirds of the incubation period.
The eggs figured in the present book were candled repeatedly for the purpose of obtaining photographs, but they nevertheless all produced healthy hatchlings in the end. This must by no means be construed as a recommendation of candling, however.

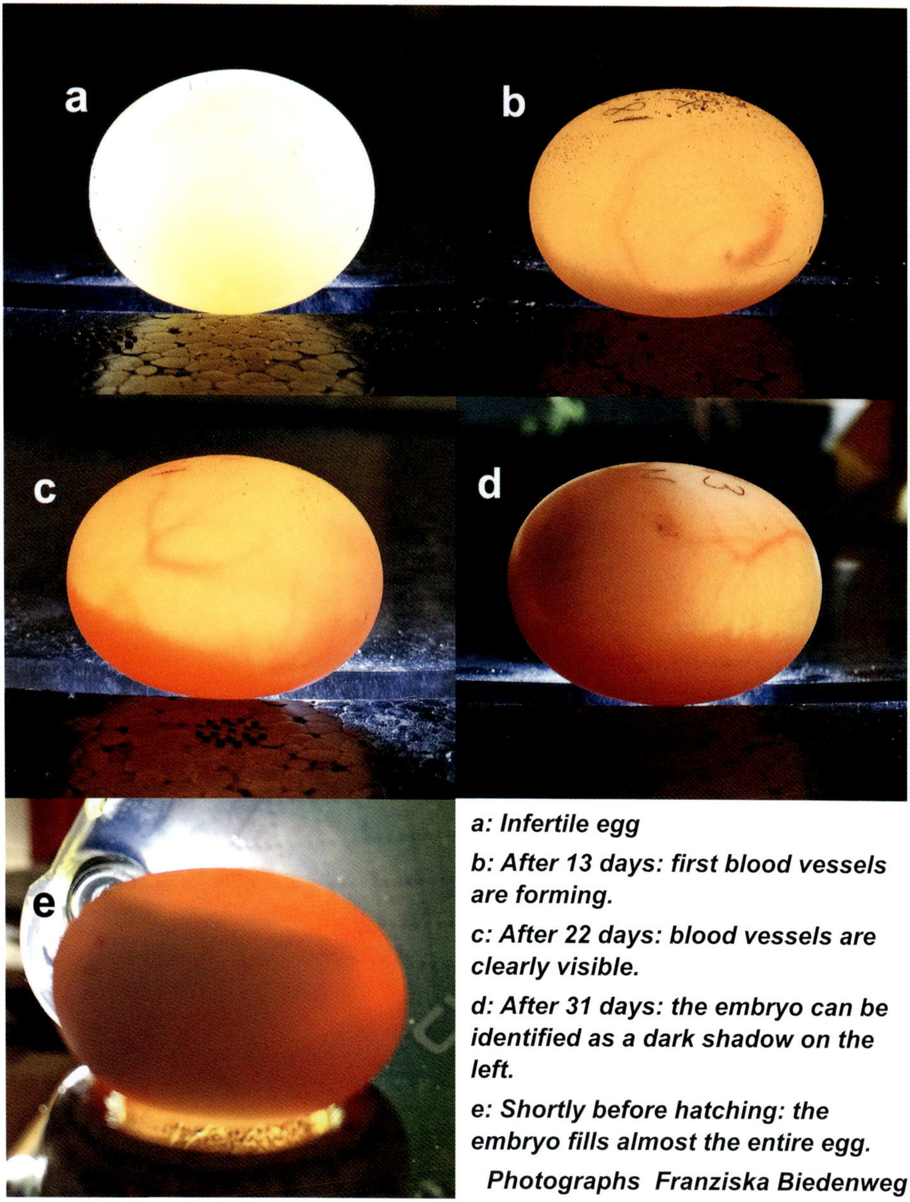

a: Infertile egg

b: After 13 days: first blood vessels are forming.

c: After 22 days: blood vessels are clearly visible.

d: After 31 days: the embryo can be identified as a dark shadow on the left.

e: Shortly before hatching: the embryo fills almost the entire egg.

Photographs Franziska Biedenweg

Hatching:

Information on incubation periods in the natural habitats is limited to one observation "after days" (Geffen & Mendelssohn 1991), and newly hatched juveniles were found in late autumn and winter (Gruber 1998).

The incubation period of eggs incubated by artificial means depends on a range of influential factors, which explains why data differ in part extensively. Most reports stated that hatching took place after about 70–90 days. An intriguing detail in this conjunction is that eggs incubated at lower temperatures with the aim of producing males apparently hatched earlier on average that those incubated at higher values (BRANSER 2011, SCHEFFEL 2014), which has remained an unexplained phenomenon thus far. Individual eggs may also take much longer to incubate, triggering discussions revolving around a possible diapause (a break in the development) that could be caused by certain outside influences, e.g., the onset of the rainy season in autumn (GEFFEN & MENDELSSOHN 1991, MENDELSSOHN 1982), and MENDELSSOHN (1982) found an embryo in an egg that was still alive after ten months of incubation.

LORTET (1887) alleged that it would only take a sufficient rise in temperature for *Testudo kleinmanni* to hatch already after twenty days, which is hardly reasonable, and that the hatchlings would be yellow throughout. PRITCHARD (1979) also mentioned that newly hatched babies would be largely yellow overall with a dark spot on each central scute. This contrasts with the majority of other reports, photographs and our own observations that the carapaces of hatchlings would be rather brownish with a marbled pattern (see p. 29) and only the plastron would be yellow with a few dark spots. It may be presumed that entirely yellow hatchlings are rare, or that there are differences between individual, geographically localised populations, or that yellow hatchlings are the offspring of rare yellow parents.

The actual hatching process begins with the embryo piercing its eggshell with a small hard protrusion on the tip of its beak, the so-called egg-tooth, which will remain present for some time after hatching. The hole created thus will subsequently be widened with the aid of the head and the front legs. It may take up to one week before the hatchling eventually exits its egg, during which period it will likely use up what is left of its yolk reserve. Once the baby has emerged from its egg, it is taken out of the incubator and checked as to whether the navel has closed properly and that no remnants of its yolk sac are left. The plastron will still sport a ventral fold right after hatching. The hatchling is then weighed and placed into a small dish with a very low level of lukewarm water. Most hatchlings will respond to this by starting to drink greedily (BIEDENWEG own obs., BRANSER 2011, VINKE & VINKE 2004). Baby Egyptian Tortoises usually weigh 4 to 6 g on hatching.

It always appears as a small miracle to see these tiny yet perfect tortoises emerge from their eggs – and it always presents the keeper of Egyptian Tortoises with great joy and a very rewarding highlight.

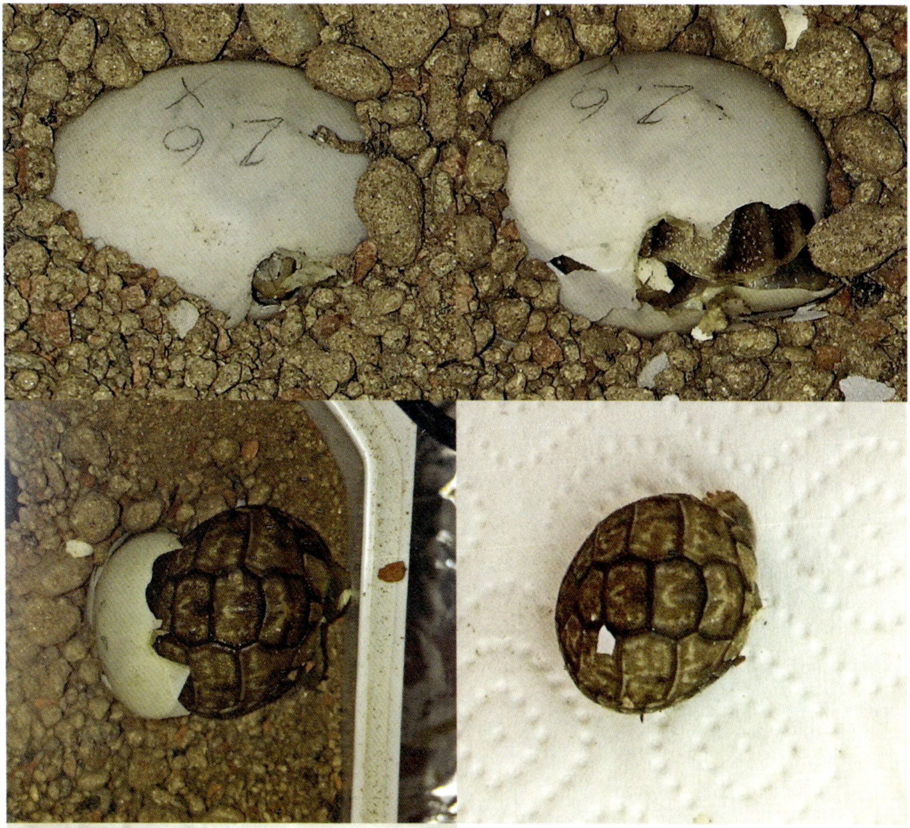

The hatching process:
- ***The egg is pierced with the egg-tooth.***
- ***The hole is widened with the aid of the front legs.***
- ***Hatching is completed.***
- ***The carapace right after hatching***
- ***The plastron with the ventral fold right after hatching***

Photographs Franziska Biedenweg

On 16 May 2018, a keeper of Egyptian Tortoises discovered twins in one of a batch of incubated eggs. While such events have been noted occasionally before in Mediterranean tortoises, this was probably the first documented hatching of twins in *Testudo kleinmanni*.

Twins in Testudo kleinmanni

Photograph Pier Giorgio Murgia

In most cases of twinned tortoises, one twin will be smaller than its sibling and unlikely to survive, and such was the case also in the instance described here. Twins share a yolk sac in their common egg and its remains need to be split in order to improve the chances that at least one will survive, if not both in the best-case scenario. This is typically performed in sterile conditions by a qualified veterinarian who will tourniquet the yolk sac with a sterile thread towards each sibling and then cut the sac between the two threads. In the case described here, the smaller twin had already perished and needed to be separated in this manner from its larger sibling. By the time of writing, the larger twin had survived the operation and was developing satisfactorily.

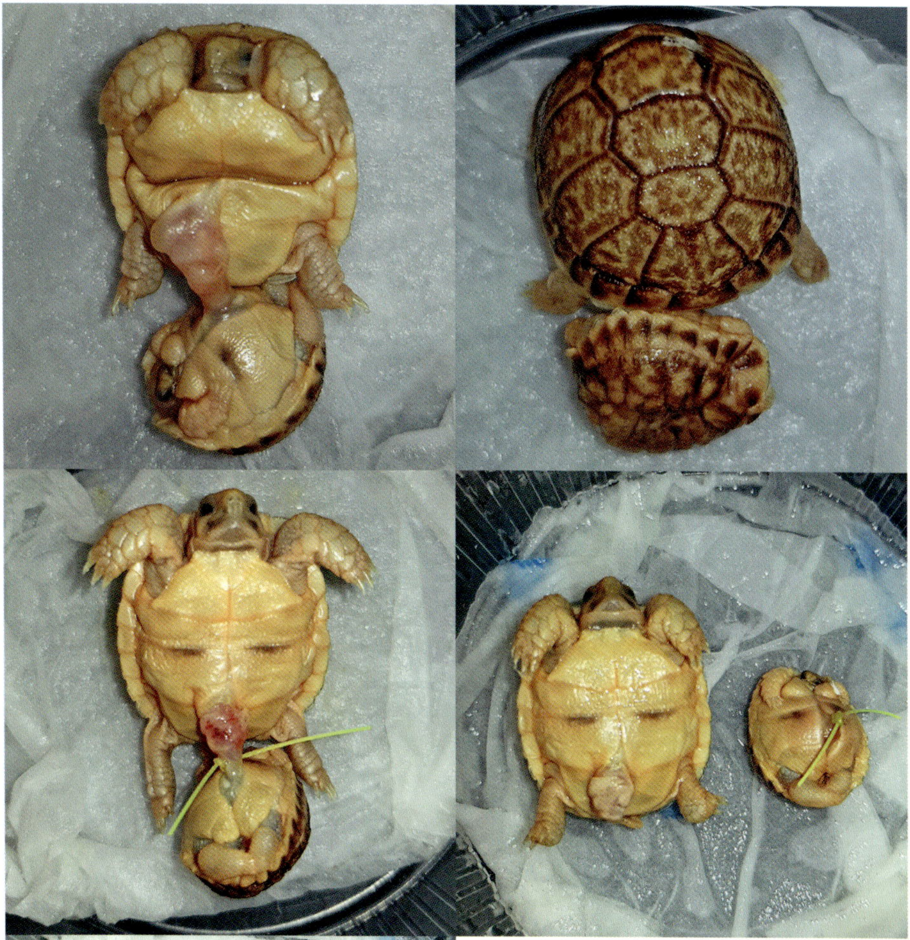

Separating the twins from one egg of an Egyptian Tortoise

Photographs Pier Giorgio Murgia

Complications: Incubation and hatching are not always straightforward affairs and complications must be expected at all times.

1. Cracks appearing in the eggshell – Causes include humidity values that are too low or too high, a moisture gradient between the egg and its environment, or the egg taking in excessive moisture (Eggenschwiler 2000). A crack does not inevitably mean that the embryo will not be able to develop further. In fact, small cracks do not appear to negatively impact on the embryonic development (Vinke & Vinke 2004). Larger cracks can be carefully resealed with the aid of soft wax (Köhler 2004, Vinke & Vinke 2004), and applying a short length of sticky tape (Branser 2011) or superglue (Scheffel 2014). for this purpose have also saved the one or other cracked tortoise egg. Cracks may of course be the result of the egg being infertile or the embryo having perished so that decomposition fumes eventually burst the egg open. If you detect the smell of decomposition, the egg giving it off needs to be removed without delay.

Cracks forming in an egg of an Egyptian Tortoise

Photograph Franziska Biedenweg

2. Damage to the embryo through light or precipitating water – Continuous bright light in the room in which the incubator is set up may cause the development of eggs to come to a halt and/or kill developing embryos (Scheffel 2014). Even more dangerous is precipitating water dripping onto the eggs, even if only very temporarily, or excessive remoistening of the incubation substrate, both of which may cause serious damage (see pp. 202, 207).

3. Appearance of mould – Causes for mould include excessive moisture and humidity, spoiled eggs, or perished embryos. Mould will typically form on infertile or

perished eggs. If it spreads over the entire egg, it will have to be removed in order to prevent the infection of neighbouring, healthy eggs. If this has already happened and is as yet limited to a small spot, an antimycotic ointment can be applied with an earbud to solve the problem (KÖHLER 2004). Mould is often noted only when a long-overdue egg is opened.

Mould growing on an infertile egg of an Egyptian Tortoise

Photograph Franziska Biedenweg

4. Perishing embryos – All the factors mentioned before may cause embryos to perish inside their eggs. Other causal factors include genetic faults and/or embryonic malformation.

A perished embryo of the Egyptian Tortoise, still with a large yolk sac attached

Photograph Michael Hebbeler

5. Premature hatching – Humidity values that are too high, excessive remoistening of the incubation substrate, and precipitating water that drips onto the eggs may all cause embryos to hatch prematurely and with a large yolk sac still attached (Jasser-Häger & Winter 2007, Vinke & Vinke 2004). If this has happened, the hatchling should be left in the incubator and kept in a separate box that is laid out with a moist kitchen towel or a sterile gauze compress until it will have resorbed the remaining yolk. In this case, the substrate may be wetted with sterile physiological saline solution rather than with water (Scheffel 2014). The moist kitchen towels or compresses need to be replaced at least once every day in order to keep the germ load in check and the risk of infection at a minimum. Only once the hatchling is stable and the navel has closed may it be transferred to a nursery terrarium (Branser 2011, Vinke & Vinke 2004, Wittmund 2005). Some authors take this procedure one step further and will first keep such a hatchling in a small nursery box that is also laid out with kitchen towels, a small shelter furnished with moist kitchen towels, and sports sources of light and warmth before it is eventually housed in a regular nursery terrarium (Scheffel 2014, Stettner 2004).

6. Anomalies – Aside from very rare instances of severe physical anomalies, such as missing eyes, malformed limbs, or stunted tails, aberrantly formed horn scutes are also rare, but still the most commonly seen anomalies. A possible cause for their appearance might again be found in excessively high incubation temperatures. A genetic predisposition has also been discussed, because offspring from parents with scutellation anomalies have been observed to more commonly feature the same anomalies in the same positions (Meyer de Rojas 2013, 2018).

A juvenile Egyptian Tortoise with a scutellation anomaly (bottom) compared to one with regularly formed scutes

Photograph Franziska Biedenweg

Scutellation anomalies are also noted in free-ranging tortoises and will not normally affect the lives of animals marked thus as long as they are not combined with other, more serious impediments. Keepers often emotionally perceive these animals as special and value their high degree of individuality.

Raising Juveniles:

Stable hatchlings with a closed navel are transferred to a nursery terrarium (see p. 136) where they will often spend the first few days buried in the moist sand. If your husbandry strategy is based on nature the hatchlings should be kept and fed just like the adults right from the beginning. This means that their nursery tank will be set up and outfitted like one for adult Egyptian Tortoises, only with everything being smaller. They will have a greater need for moisture initially, which necessitates that the bottom substrate be sprayed more regularly and they be bathed every now and then in addition. Of utmost importance is supplying them with adequate amounts of calcium and UV radiation.

Their growth should be monitored by weighing them at regular intervals and documenting their gains for comparison with their weight at hatching.

Juvenile Egyptian Tortoises gain weight more rapidly than Mediterranean tortoises. They may thus reach 1.5 times their weight at hatching after three months, three times as much after six months, and five times as much or even more after twelve months (Biedenweg own obs., Branser 2011a, b, Gad 2000, Fritzsche & Fritzsche 2002, Scheffel 2014, Wittmund 2005). Growth is

Weight check of a seven-day old Testudo kleinmanni.

Weight at hatching: 5.11 g
Weight gained: 0.82 g

Photograph Franziska Biedenweg

a matter of individuality and can vary significantly, however, with differences possibly being due to sexual identities (BIEDENWEG own obs., SCHEFFEL 2014).

The shell of a six-month old *Testudo kleinmanni* will be notably harder than that of a Mediterranean tortoise of the same age in which it is still rather elastic. The reason for this being so could be the architecture of the characteristically thicker bone shell of Egyptian Tortoises.

The conditions leading up to a period of summer dormancy are created right from the beginning, even though juveniles will remain more active during this period for their first two years, or even three in some instances, than adult *Testudo kleinmanni* and start observing it more strictly only thereafter (BIEDENWEG own obs., SCHEFFEL 2014). Only if hatching took place shortly before the time of aestivation arrives, which would probably never happen in the wild, hatchlings cannot be subjected to it right away, but must first be kept in spring conditions and supplied with fresh feed and water.

A hatchling Testudo kleinmanni in its nursery terrarium

Photograph Franziska Biedenweg

Health Issues

One important statement must be made right at the beginning here: The following chapter cannot by any means replace your consulting with a qualified veterinarian and his or her designing a therapy plan if your tortoises run into trouble. It is much rather intended to serve as a catalogue of which symptoms and changes should be looked out for and will make a visit to the vet necessary if they are noted.

In this conjunction it is highly recommendable to locate a veterinarian versed in diagnosing and treating reptiles even before you set out to acquire your Egyptian Tortoises and you do not actually need him or her right now. Vets with an advanced education in matters reptilian are not particularly common and you may find that the one nearest to you is actually quite some travelling distance away. Directories of vets specialising in this field are published and maintained on the Internet, however, and the larger terrarium associations and societies of most countries at least should be able to help if you have problems locating one.

Healthy tortoises: In order to be able to detect disease-induced changes in the appearance and behaviour of a tortoise you will obviously need to know what makes up a healthy tortoise, be entirely familiar with the specimens in your care, and have an eye on them at all times.

Traits of a healthy Egyptian Tortoise are:

- Open, clear, shiny eyes with no tear flow, not sunk in, eyelids not thickened or swollen;
- Nostrils dry and unobstructed;
- Skin of limbs, head and tail dry and not oozing liquid, no larger patches of skin detaching (only very small spots will come loose during a moult);
- Shell firm and smoothly grown;
- Walking with visible vigour with the shell well raised off the ground;
- Very active;
- Feeding with a visible appetite.

Very many health issues are unfortunately the result of flawed husbandry conditions. The consequences of compromised husbandry parameters on the health of captive Egyptian Tortoises have been outlined already throughout the preceding chapters so that we will discuss only a few additional examples here.

Humped growth: Humped growth is by no means just a cosmetic fault in the shape of pyramidal horn scutes, but the externally visible symptom of an underlying growth problem. The bone shell beneath will be spongy, thickened and deformed, and if the humped growth is massive, the volume of the body cavity may be compromised to such an extent that secondary problems will be encountered.

Causes for such a growth anomaly include a diet that is too rich in proteins and too poor in fibres, is deficient in calcium and/or vitamin D_3, or that the exposure to UV radiation is inadequate, or that the aestivation period was skipped and caused an unnaturally accelerated growth. It may be just one of these factors or several combined (Eggenschwiler 2000, Highfield 2010, Kölle 2009, Wiesner & Iben 2003), and juveniles respond significantly worse to them than adult specimens due to their growing faster and their bone tissues not yet being completely solidified.

Once the symptoms of humped growth have become manifest, not even the best husbandry conditions will be able to reverse them.

Slightly humped shell growth in Testudo kleinmanni, presumably caused by husbandry conditions that were too dry during the first months of life

Photograph Franziska Biedenweg

Softening of the shell, rickets, osteomalacia: In an extreme scenario, the carapace is sunken in, but in any case will feel soft and rubbery to the touch, and distinctly yield to pressure.

Some of factors listed under humped growth above may also cause a softening shell: calcium deficiency, UV deficiency, insufficient vitamin D_3 or its synthesis, and unnaturally accelerated growth, added to which can be kidney diseases, parasitoses, or chronic infections.

Bladder stones, urolithiasis: Urinary stones may form both in the bladder and in the urinary tracts. Symptoms are uncommon until, sometimes, the hind legs show signs of paralysis or solid constituents are noted in the urine, such as gravel or small stones. If the latter are found chances will be that larger stones that cannot be expulsed freely are also present and this would necessitate a visit to the vet without delay.

Causes are presumed to include a diet too rich in protein, an environment that is too dry, insufficient opportunities to drink, but also an inadequate supply with vitamins and calcium.

Bladder stones in Testudo kleinmanni

Left: Expulsion from the cloaca ***Right: A dropped bladder stone***

Photographs Joe Stammet

Aside from husbandry-induced diseases, infections may be imported from outside sources, and we will provide a few examples here without a claim to completeness.

Intestinal parasites: Infections with intestinal parasites cannot be prevented, as these will be imported all the time with fresh feeder plants from outdoor sources. They can furthermore be transferred from one tortoise to the next via faeces or during mating. They are basically harmless, but may proliferate massively in situations of stress. If any symptoms or suspicious behavioural changes suggest you had better go and see a vet, you should always take a faecal sample along for a massive infection with intestinal parasites could be at the root of the problem.

Tortoises are known to be carriers of pathogenic parasites such as oxyurans, ascarids, certain flagellates, more rarely amoebas and ciliates. Prophylactic deworming without prior faecal examination must be discouraged because, for one, a possible infection with pathogenic unicellular organisms that may need to be treated will not have been established, and, for the other, the organism of the tortoise will be burdened with pharmaceuticals that may not even be needed.

A massive infection with oxyurans, macroscopically visible in the faeces of a Testudo kleinmanni.

Photograph Franziska Biedenweg

It is therefore much rather advisable to have a veterinarian versed in treating reptile patients check faecal samples routinely once every year. The most suitable time for doing so is at some point during the activity period of Egyptian Tortoises, but at least six weeks before the start of their summer dormancy period so that there will be enough time if it turns out to be necessary to deworm them and subsequently have the residues of the medication break down and be flushed out of their system before they start aestivating.

Faecal samples should be as fresh as possible when they reach the vet. It is not really necessary to have each and every individual's faeces checked. A collective sample will do, because if it turns out positive for the one or other parasite the entire colony will have to be medicated anyway. To this end, you will need to collect as many fresh droppings as possible and store them, e.g., in a sample tube from the pharmacy or simply in a watertight vessel with a tightly closing lid after adding a few drops of water to keep them hydrated. If they are not taken to the vet right away they must be kept refrigerated.
Simple hygienic measures like new acquisitions in quarantine, disinfecting hands between effecting works in different terraria, removing feeding and water bowls daily, and collecting faeces and soiled leftovers of food on a daily basis will all contribute to keeping intestinal parasites from proliferating excessively.

Viral diseases: Just about every keeper of tortoises will have heard or read about infections with herpes viruses. They are greatly feared, because they will almost always have a deadly outcome earlier or later. By now, several species of herpes viruses and other pathogenic viruses have been identified in tortoises and become identifiable in pathology labs. Not all tortoise species respond equally promptly and with severe symptoms to an infection with herpes, and Marginated Tortoises (*Testudo marginata*), Spur-thighed Tortoises (*Testudo graeca*) and Central Asian Tortoises (*Testudo (Agrionemys) horsfieldii*), for example, may continue to live for a very long time without showing signs of being infected – and spread the virus to other specimens all the while. As far as *Testudo kleinmanni* is concerned, such an infection will progress relatively rapidly and soon see affected specimens perish, however, as has been observed in smuggled and confiscated specimens (Marschang et al. 2009).
Symptoms are not always evident and tend to surface only during an acute episode of the disease. Classical symptoms include stomatitis (an inflammation of the oral mucous membranes) with yellowish deposits on the tongue, rhinitis ("runny nose syndrome"), blepharo-oedema (swelling of the eyelids), and possibly pneumonia with associated breathing difficulties. An acute episode with symptoms is often triggered by stress such as major changes in the environment or suboptimal husbandry conditions. The viruses are transferred from one animal to another via saliva, nasal secretions, faeces and possibly also sperm. There is no cure for such an infection and treatment will be limited to mitigating the symptoms with medication.
Diagnosis is possible only during an episode with acute symptoms when a swab from the tongue will enable the pathology lab to find the virus. A blood sample may also provide clues via an antibody titre estimation during latent stages, but only if the infection has been present for a rather long time and antibodies

have been developed. It has therefore been suggested to have blood tests for herpes viruses done repeatedly at intervals of at least six weeks if a tortoise is suspected to possibly be infected (EGGENSCHWILER 2000, KÖLLE 2009). Strict quarantine, which is discussed in greater detail below, will be necessary in such a situation.

Mycobacteria: Mycobacteria or mycoplasma are minuscule bacteria without cellular walls that cause numerous diseases in both humans and animals. In tortoises, they will be at the root of the runny nose syndrome (RNS) or URTD (upper respiratory tract disease). Transmission takes place via a smear infection from an infected animal to another. Symptoms include the formation of blisters at the nostrils and a watery discharge from the nose, escalating into viscous slime obstructing breathing when it will have turned chronic. Mediterranean tortoises kept under optimal husbandry conditions will often go without symptoms and be affected only in rare instances. Acute episodes may then be caused by stressful situations such as changes in the immediate surroundings, other diseases, or suboptimal husbandry conditions like low temperatures or inadequate moisture/ humidity levels.

In *Testudo kleinmanni*, like in various exotic tortoise species, the course of such an infection may be more serious. Affected specimens will additionally exhibit a clearly compromised general health status with disinterest in food and lethargy. Accompanying symptoms may include conjunctivitis ("pink eye"), blepharitis (swelling of the eyelids), and sunken eyes. Secretions may collect in the lungs and cause severe pneumonia, notable from rasping breathing sounds and laboured breathing (JACOBSON et al. 1991, SCHUMACHER et al. 1993, BLAHAK 2009b).

This infection can be identified by examinations of a sample obtained through nasal flushing or tests for antibodies from a blood sample.

An mycobacterial infection is not completely curable either, but specialised antibiotics will be able to keep symptoms in check and mitigate them during an acute episode.

Additional information on the mentioned and other diseases can be found in specialist subject literature, (e.g., BLAHAK 2009a, b, 2013, BRAUNE et al. 1989, BROSDA 2011, BROWN et al. 2004, CLAUSSEN 1981, CORTON 1996, EGGENSCHWILER 2000, FRYE 1991, GIANETTO et al. 2007, HÄFELI & ZWART 2000, HIGHFIELD 1998, 2003, 2005, 2010, JACOBSON et al. 1991, 2014, KÖLLE 2000, 2009, KÖLLE & BLAHAK 2015, KRUG 2010, MARSCHANG 2011, MARSCHANG et al. 1997, 2001, 2009, MARSCHANG & SCHNEIDER 2007, MATHES 2003, REINAUER et al. 2005, SASSENBURG 2005, SCHUMACHER et al. 1993, UNE et al. 1999).

A veterinarian administering medication to a Testudo kleinmanni

Photograph Franziska Biedenweg

In the case of a health problem that requires medicating, the vet will explain why and how medication needs to be administered and, if necessary, demonstrate the process.

Quarantine: One of the most important measures to keep an existing tortoise colony healthy is to strictly quarantine new acquisitions. Quarantine should be continued for at least one year during which period several tests need to be conducted to ensure that the new specimen will definitely be free of herpes viruses and other potentially devastating diseases. Faecal samples should furthermore be checked for parasites twice during this time. Here it will not suffice to partition off a section in an occupied terrarium – you need a separate tank. Using one-way gloves and thoroughly disinfecting hands are measures you will have to become used to, as is taking care of the quarantined specimen(s) last in your routine.

Strict quarantine is particularly critical in the case of new acquisitions of adult Egyptian Tortoises or those of questionable origin. Even juveniles from different captive-bred stocks should be kept separated, monitored for some time, and their faeces checked for parasites, even if blood samples for herpes tests cannot be taken due to their small body volumes.

Legal Considerations

The Convention on International Trade in Endangered Species of Wild Fauna and Flora (CITES) controls the international trade in animal and plant species taken from the wild and constitutes one of the most important international agreements on the protection of natural resources. In it, *Testudo kleinmanni* has been included in Appendix I since 1994, which is the highest protection category. It means that this species is considered particularly seriously endangered and immediately threatened with extinction. The commercial trade in wild-caught specimens of these species is therefore outlawed. The enforcement of the CITES convention, including the trade in specimens produced in captivity, is regulated and controlled by the individual signatory states.

Within the European Union (EU), the EU Ordinance on Species Protection applies. Here, *Testudo kleinmanni* is listed in Annexure A, which includes all species listed in CITES Appendix I as well as a few additional ones. Again, the individual member states of the EU are entrusted with enforcing the ordinance, which may in some instances have been delegated further to their provincial or federal state administrations.

What does this mean within the EU?

- No Egyptian Tortoises collected from nature may be imported or traded.
- Specimens may only be traded or passed on otherwise if they originate from legal parents.
- Each and every *Testudo kleinmanni* must have been issued an EU-certificate with a trade permit and be individually identifiable by means of photographic documentation or, in some countries, be furnished with a transponder chip. Breeders will therefore have to apply for these documents by submitting sets of initial photographs, as may be required for example, to the authority that handles these procedures in their countries. This certificate will then be valid throughout the EU so that specimens may also be taken or send across the national borders between EU member states.
- Registration is obligatory, with keepers and breeders having to notify the relevant authority of every single specimen newly obtained or hatched or passed on to another owner. The new owner will then have to register his or her new acquisition(s) in turn. This procedure also applies if the owner of the subject tortoises moves residence or after an animal has perished.

Laws and procedures may differ from one country to the next or even from one province or state to the next. It is therefore critical that every owner of Egyptian Tortoises, potential or factual, familiarise themselves with what exactly is required and which procedures are to be followed at their place of residence, preferably by visiting in person the authority charged with their enforcement. Which one that exactly is can usually be learnt without much effort from the Internet or the local terrarium keepers association.

If photographic documentation is required, photos will have to be retaken at the prescribed regular intervals and kept, dated and furnished with the weight of the specimen at the time the shot was taken, with the respective EU certificate (Bender 2001, Bender et al. 2007), failing to do so will void it and the animals may be confiscated by the relevant authority. No such updating is required in the case of an implanted transponder chip, of course, whose ID number will be shown in the certificate.

If an Egyptian Tortoise is to be exported to a non-EU country, the seller will have to apply for a CITES export permit. This is usually done at a higher-level office and by means of a form that must be authenticated by the authority at which the specimen is registered. The purchaser will in turn have to apply for a CITES import permit at the authority handling these matters in his or her country. Aside from all these, certain animal welfare regulations may apply that must be adhered to just as strictly, information about which can usually be obtained from the state veterinary services office. A trans-border transfer of ownership may furthermore necessitate involving customs, the state or border veterinarian, or other authorities of either or both the involved countries. Here, you will have to find out what exactly is prescribed.

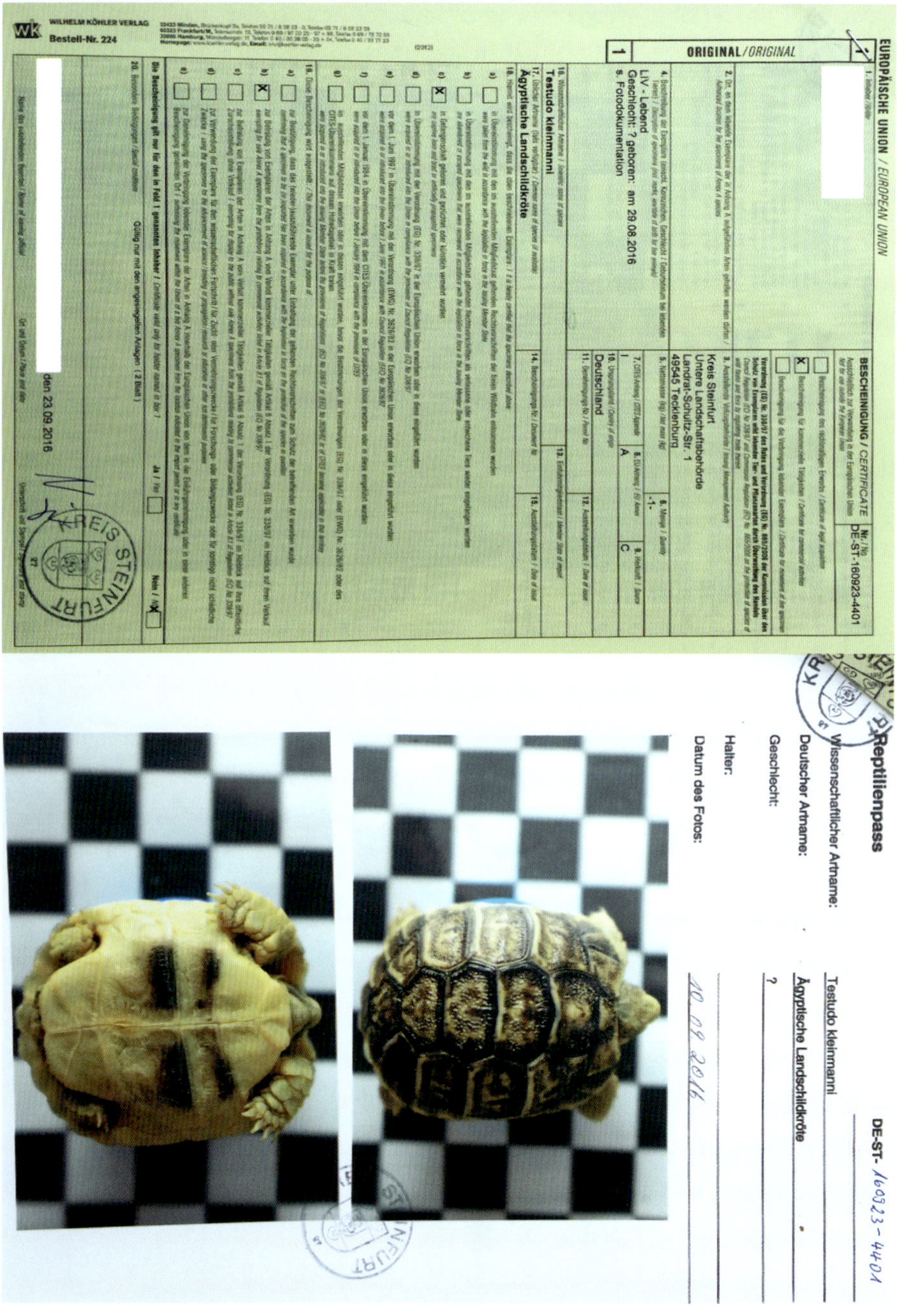

A EU-certificate and the photographic documentation for a Testudo kleinmanni **Photographs Franziska Biedenweg**

A juvenile Testudo kleinmanni
Photograph Franziska Biedenweg

Before and After Acquisition

Now the time has arrived – you have informed yourself comprehensively and everything has been set up and tested so that some Egyptian Tortoises can finally move in with you. The best time for that is obviously during the activity phase of these tortoises in late autumn or winter.

As has been mentioned before, juveniles tend to adjust to their new surroundings much more readily than adults. The former are most commonly acquired from a breeder whom you will have found via a classified ad on the Internet. Once contact with him or her has been established, you may want to ask questions about how to best accommodate the animals after their arrival. As yet, there are not too many breeders of this species so that you may be required to travel some longer distance. It will but in any case be of great advantage if you make the effort and so be able to obtain a personal impression of the breeder, his or her setup, and have your last questions answered. Most breeders will likewise prefer meeting the new owner face to face to assure themselves that their baby tortoises go to a good new home and are looked after properly in the future as well. If you do not want to travel the distance repeatedly and want to take your juveniles home right away, everything needs to be in place before you leave. The terrarium must be completed in all details, the bottom substrate must have sufficiently dried, the associated technology must be installed, and everything should have been tested and running to perfection for some time.

If the opportunity presents itself, you may want to make contact with several breeders, form an opinion on each of them, and eventually choose one of them. A breeder worthy of your trust will:

- answer all your questions and give advice on husbandry, technological aids, and outfitting;
- offer to tutor you after you have made your purchase;
- insist on assuring him/herself that you provide the animals with optimal care;
- readily show you their terraria or pens and the parent specimens of your juveniles when you visit them.

During your visit to the breeder, you should pay attention to the following details:

- The Egyptian Tortoises live in adequately large terraria or indoor pens.
- The outfitting enables the tortoises to live according to their biological needs – bright light, UV lamps, adequate temperatures and humidity, structured decoration, etc..
- All tortoises look healthy (see p. 223).

- The juveniles are active and free of suspicious signs – the shell is hard, the eyes clear, the nose dry, they walk on stilted legs, etc..
- The required original EU-certificates are ready and complete with the necessary photographic documentation that will let you identify every individual without doubt, or a scan of the implanted microchip matches the number given in the document.

One question keeps on popping up when it comes to keeping tortoises: Should you keep several individuals together or rather individually? Opinions differ on a correct answer and tortoises are often considered individualists. Field observations have clearly revealed that the juveniles of Mediterranean tortoises will stay together in colonies for their first few years of life (Wegehaupt 2006, 2009, 2012, 2019 b), because a group affords better protection against predators. It is also during this period that the individual members will be imprinted on their future mating partners. Even though no social behaviour in a stricter sense may be obvious, adult tortoises will interact all the time with their co-inhabitants of the same territory, are familiar with these, sometimes share their shelters with them, and are able to identify alien specimens immediately. The keeper will therefore find it much more interesting to watch a colony of tortoises for they are much more active and feed with a greater appetite than individually kept ones. Breeders will therefore usually part with their captive-bred juveniles only at least as pairs or for adding individuals to an existing colony.

Transport is best effected using a cooler or polystyrene box, especially considering that it will typically take place in the inclement weather of winter. The tortoises themselves should be accommodated in individual plastic boxes, which will keep them from scratching at the polystyrene walls and provide enough space for adding a small bottle with warm water if the day is particularly cold. The boxes are furnished with slightly moistened kitchen towels and some hay. Soil or sand are not suited in this instance as they would stick to the animals. The stress of shocks and vibrations during transport will usually cause the tortoises to completely empty their bowels and bladder, which causes them to dehydrate. This must be compensated for immediately on your arrival home by affording them a bath in very shallow, lukewarm water. Once they have soaked and drunk enough they can be released in their new home. They must be left completely alone and not be handled over the next days unless absolutely necessary. Their adjusting to the unfamiliar environment will be all the easier for them if the ambient parameters about correspond to what they were used to at their breeder's place.

Photograph Franziska Biedenweg

We have come to the end of our book and say Good-bye. We thank you for your attention and hope you have enjoyed reading it. Perhaps it has managed to attract your interest and maybe even enthusiasm for a fascinating little beauty called *Testudo kleinmanni*.

Questions, suggestions, constructive criticism, but also positive comments are very welcome and may be directed to info@tartaruga-verlag.de.

We wish you lots of enjoyment in your keeping and watching your *Testudo kleinmanni* and sign off,

Franziska Biedenweg and Ricarda Schramm

Acknowledgements

We wish to express our heartfelt gratitude to Aviad Bar, Michael Hebbeler, Andreas Krb (Reptiles Expert! GmbH Austria), Pier Giorgio Murgia, Joe Stammet, and The Libyan Wildlife Trust for their making available photographic material.

Thanks are also due to Martina and Sven and their family for their being so accommodating and understanding.

Not least do we thank our life partners, Marvin Fuchs and Rudolf Schramm, respectively, for their patience and support during the creation stage of the present book.

An additional personal Thank-you goes to Rudolf Schramm for his advice and assistance with the construction of terraria and solving technological problems.

References

Acierno, M. J., M. A. Mitchel, M. K. Roundtree & T. T. Zachariah (2006): Effects of ultraviolet radiation on 25-hydoxy-vitamin D(3) synthesis in red-eared slider turtles (*Trachemys scripta elegans*). – American Journal of Veterinary Research 67(12): 2046-2049

—, M. A. Mitchel, M. S. Kirchgessner & D. Sanchez-Migallon Guzman (2008): Effects of ultraviolet radiation on plasma 25-hydroxyvitamin D3 concentrations in corn snakes (*Elaphe guttata*). – American Journal of Veterinary Research 69(2): 294-297

Akeret, B. (2013): Energiewende im Terrarium. Ideen und Vorschläge, wie man Stromverbrauch und Energiekosten bei der Haltung von Terrarientieren senken kann. – REPTILIA 99, 18(1): 18–31

— (2015): Pflanzen im Terrarium. Anleitung zur Pflege von Terrarienpflanzen, zur Gestaltung naturnaher Terrarien und Auswahl geeigneter Pflanzenarten. – Natur und Tier-Verlag, Münster, 408 S.

— (2017): Pflanzenpflege in Pfeilgiftfroschterrarien – Grundsätzliche Überlegungen. – REPTILIA 123, 22(1): 26–31

Allen, M. E., T. C. Chen, M. F. Holick & E. Merkel (1998): Evaluation of Vitamin D status of the green iguana (*Iguana iguana*): oral administration vs UVB exposure. – In: Holick, M. F. & E. G. Jung (Hrsg.): Biologic effects of light (1998). Kluwer, Boston: 99-102 (https://page-one.live.cf.public.springer.com/pdf/preview/10.1007/978-1-4615-5051-8_17)

Ammermüller, J. & Kolb, H. (1996): Functional Architecture of the Turtle Retina. – Progress in Retinal and Eye Research, 15(2): 393–433

Anderson, J. (1896): A Contribution to the Herpetology of Arabia. With a Preliminary List of the Reptiles and Batrachians of Egypt. R. H. Porter (Ed.). – Taylor & Francis, London: 126 S.

— (1898): Zoology of Egypt, REPTILIA and Batrachia, Vol. 1. – Bernard Quaritch, London: 28–31

Andersson, L. G. (1903): List of reptiles and batrachians collected by the Swedish Zoological Expedition to Egypt, the Soudan and the Sinaitic Peninsula. L.A. Jägerskiöld Results of the Swedish Zoological Expedition to Egypt and the White Nile, 1(4): 1–12

Anon (1985): *Testudo kleinmanni* in Egypt and Israel – IUCN Tortoise Group Newsletter (Maitland/Florida), No. 1:3

Antenbrink-Vetter, S. & Vetter, H. (1995): Schildkröten-Lexikon: Ägyptische Landschildkröte *Testudo kleinmanni* Lortet, 1883 – Schildkröten 2(3)

— & Vetter, H. (1999): Die Schildkröten Ägyptens – Schildkröten 6(1): 4–7

Arbel, A. (1990): *Testudo kleinmanni*. In: Plants and Animals of the Land of Israel, an Illustrated Encyclopedia, Volume 5 Reptiles and Amphibians, Alon, Azaria (Ed.). – Ministry of Defence Publishing House/ Society for the Protection of Nature in Israel, Jerusalem: 37–38

Arnold, K. E. (1986): Untersuchungen zur Farbwahrnehmung der Rotwangenschmuckschildkröte *Pseudemys scripta elegans*. – Unpublished thesis, Johannes Gutenberg Universität, Mainz: 130 S.

— & Neumeyer, C. (1987): Wavelength discrimination in the turtle *Pseudemys scripta elegans*. – Vision Research, 27(9): 1501–1511.

Arth, S. (2012): Die Leuchtstoffröhre – ein Klassiker neu beleuchtet. – TERRARIA/elaphe 5/2012: 18–25

Attum, O. (1997): A note of the sex ratio of the Egyptian tortoise, *Testudo kleinmanni*; British Herpetological Society Bulletin 61: 14–15

—, S. Baha El Din, S. Carranza, R. Earley, E. N. Arnold & B. Kingsbury (2007): An evaluation of the taxonomic validity of *Testudo werneri*. – Amphibia-REPTILIA 28(3): 393–401

—, S. Baha El Din, W. E. Farag, M. M. Esawy, A. E. Gad & B. Kingsbury (2007): Returning them back into the wild: Movement patterns of repatriated Egyptian Tortoises, *Testudo kleinmanni* Lortet, 1883 (Sauropsida: Testudinidae). – Zoology in the Middle East, 41(1): 35–40

—, M. Baha El Din, S. Baha El Din & S. Habinan (2007): Egyptian tortoise conservation: A community-based, field research program developed from a study on a captive population. – Zoo Biol. 26(5): 397–406

—, B. Rabea, S. Osman, S. Habinan, S. M. Baha el Din & B. Kingsbury (2008): Conserving and studying tortoises: A local community visual-tracking or radio-tracking approach? – Journal of arid Environments 72(5): 671-676.

—, W. E. Farag, S. M. Baha El Din & B. Kingsbury (2010): Retention rate of hard-released translocated Egyptian tortoises *Testudo kleinmanni*. – Endangered Species Research, Vol. 12: 11–15

—, R. Bassim, K. Duffy & M. Baha El Din (2011): Testing the reliability of ring counts for age determination in the Egyptian tortoise (*Testudo kleinmanni*). – Herpetological Journal 21(3): 209–211

—, A. Kramer & S. M. Baha El Din (2013): Thermal utility of desert vegetation for the Egyptian tortoise and its conservation Implications. – Journal of Arid Environments 96: 73–79

— & C. D. Cutshall (2015): Movement of translocated turtles according to translocation method and habitat structure. – Restoration Ecology 23(5): 588–594

— & B. Rabia (2016): Movement patterns of soft released translocated Egyptian Tortoises. – Journal of Arid Environments 134: 62–65

Baha El Din S. & M. Baha El Din (1992): Notes on the herpetology of North Sinai – British Herpetological Society Bulletin (41): 9–11

— & M. Baha El Din (1994): Status of the Egyptian Tortoise in Egypt. – TortoiseTrust Website. https://www.tortoisetrust.org/articles/toreport.html

— (2002): *Testudo kleinmanni* In Libya. – TortoiseTrust Website. http://www.tortoisetrust.org/articles/libya.htm

—, O. Attum & M. Baha El Din (2003): Status of *Testudo kleinmanni* and *T. werneri* in Egypt – Chelonian Conservation & Biology 4(3): 648–655

— (2006): A Guide to the Reptiles and Amphibians of Egypt. – The American University in Cairo Press. Cairo-New York: 359 S.

Baines, F. (2008): Photo-Kerato-Conjunctivitis in Reptiles. Autumn Meeting. Nov. 2008. – Stretton, Cheshire, United Kingdom. Proc. Br. Vet. Zool. Soc. P 43.

— (2010): Photo-Kerato-Conjunctivitis in Reptiles.In: 1st International Conference on Reptile and Amphibian Medicine, March 4th-7th 2010: Proceedings, Sabine Öfner & Friederike Weinzierl (Hrsg.). – Verlag Dr. Hut, München: 141-145

—, J. Chattell, J. Dale, D. Garrick, I. Gill, M. Goetz, T. Skelton & M. Swatman (2016): How much UV-B does my reptile need? The UV-Tool, a guide to the selection of UV lighting for reptiles and amphibians in captivity. – Journal of Zoo and Aquarium Research 4(1): 42–63 http://www.jzar.org/jzar/article/view/150

Bauer, A. M., J. C. DeBoer & D.J. Taylor (2017): Atlas of the reptiles of Libya. – Proceedings of the California Academy of Sciences, Series 4, 64(8): 155–318

Beetz, H. (1965): "Sie haben ja eine *Testudo kleinmanni!"*. Aquarien und Terrarien 12(6): 206

Bender, C. (2001): Photographdokumentation von geschützten Reptilien, Herausgegeben von der Deutschen Gesellschaft für Herpetologie und Terrarienkunde: 28 S.

—, K. Henle & P. M. Kornacker (2007): Standards für die Photographdokumentation von Jungtieren der Landschildkröten-Gattung *Testudo*. Nat. Landschaft 82(1): 11–19

Bidmon, H.-J. & Jennemann, G. (2006): Hohe relative Luftfeuchtigkeit – gleich glatte Panzer: Wie lässt sich das in der Landschildkrötenhaltung praktikabel realisieren? Schildkröten im Fokus 3(4): 3–18

— (2006): Die Aufzucht und Ernährung europäischer Landschildkröten – Grundlagen und Rezepte, Futtermittel und Zusatzstoffe. S. 117 – 136 in: Daubner, M. & Vinke, T. (Hrsg.): Testudo – häufig gehaltene Arten – Schildkröten im Fokus Sonderband. – Dauvi-Verlag, Bergheim.

— (2009): Ernährungsgrundlagen und Darmpassagezeiten bei herbivoren Landschildkröten – oder wie selektierende Nahrungsgeneralisten auch unter extremen Bedingungen überleben: Eine Übersicht. – Schildkröten im Fokus, Bergheim 6(1): 3–26.

— (2011): Beleuchtungsfehler bei der Haltung tropischer Landschildkröten. – Schildkröten im Fokus 8(1): 31–35

Blahak, S. (2009a): Virale Erkrankungen bei Schildkröten und deren Prophylaxe. Chem. Vet. Untersuchungsamt OWL – PDF-Download: Website http://www.schildkroeten-stammtisch-rhein-main.de/index.php/grundlagen/ersteschildkroete/cat_view/45-?start=10

— (2009b): Mykoplasmen bei Landschildkröten. Chem. Vet. Untersuchungsamt OWL – PDF-Download: Website http://www.schildkroeten-stammtisch-rhein-main.de/index.php/grundlagen/ersteschildkroete/cat_view/45-?start=10

— (2013): Virusinfektionen bei Landschildkröten – Todesfälle im Schildkrötengehege. – Hundkatzepferd 4(13): 36–39

Branser, S. (2011a): Langjährige Erfahrungen bei der Haltung und Vermehrung der Ägyptischen Landschildkröte, *Testudo kleinmanni*. – RADIATA 20(3): 2–32

— (2011b): Langjährige Erfahrungen bei der Haltung und Vermehrung der Ägyptischen Landschildkröte, *Testudo kleinmanni* – Vorgeschichte zum gleichnamigen Artikel in der RADIATA 3/2011. – MINOR 10(3): 22–26

Braune, S., V. Geiss & W. Thiel (1989): Eine neue durch Herpesviren verursachte Erkrankung bei Landschildkröten. – Tierärztliche Praxis 17(4): 416–419

Bringsøe, H. & J. R. Buskirk (1998): Distribution of *Testudo kleinmanni* Lortet, 1883 and *Testudo graeca* Linnaeus, 1758 in the Negev Desert, southern Israel (REPTILIA: Testudines: Testudinidae). – Faun. Abh. Mus. Tierk. Dresden 21(4): 23-30

Brosda, A. (2011): Untersuchungen zur Infektion mit Oxyuren bei mediterranen Landschildkröten in menschlicher Obhut und ihr Einfluss auf die Entwicklung juveniler *Testudo graeca*. – Mensch und Buch Verlag, Berlin: 163 S.

Brown, D. R., E. R. Jacobson & M. B. Brown (2004): *Mycoplasma testudineum* sp. nov., from a desert tortoise (*Gopherus agassizii*) with upper respiratory tract disease. – International Journal of Systematic and Evolutionary Microbiology 54: 1527–1529

Bruekers, J. (1994): Wetenswaardigheden over de Egyptische landschildpad. – De Schildpad 20(3): 27–42

— (1995): De Egyptische landschildpad (*Testudo kleinmanni*) – In: W. van Bakel, J. Bruekers, R. de Bruin, M. Kelderman & I. Wieberdink (Eds.): Special ter gelegenheid van het 20-jarig bestaan van de Nederlandse Schildpadden Vereniging. Uitgave van de Nederlandse Schildpadden Vereniging.

— (2001): De Egyptische landschildpad (*Testudo kleinmanni*). – Website Nederlandse Schildpadden Vereniging, 6 pages

Budde, H. (1980): Verbesserter Brutbehälter zur Zeitigung von Schildkröten-Gelegen. – SALAMANDRA 16(3): 177–180

Buskirk, J. (1985): The endangered Egyptian tortoise *Testudo kleinmanni*: Status in Egypt and Israel; In: McKeown, Caparoso und Peterson (Hrsg.): Ninth International Herpetological Symposium on Captive Propagation and Husbandry, San Diego: 35–52

— (1990): The Plight of the Egyptian tortoise. – Tortuga Gazette 26(6): 7–8

— (1993): An annotated bibliography of the Egyptian tortoise, *Testudo kleinmanni* LORTET, 1883. – Tortuga Gazette 29(1): 1–4

— (1994): Eine Bibliographie über die Ägyptische landschildkröte, *Testudo kleinmanni* (LORTET, 1883). – SIGS-Info, Juni, 3–5

— (1997): The Plight of the Egyptian tortoise in Israel. – Tortoise Trust Newsletter 12(2): 15

CALABRESI, E. (1923): Missione zoologica del Dr. E. Festa in Cirenaica. Boll. Mus. Zool. Anat. Comp. R. Univ. Torino 38: 7–8

CHAMBAL, S. (1993): Etológia, biológia a chov v teráriu korytnacky egyptskej *Pseudotestudo (Testudo) kleinmanni*, LORTET,1883. – Chamaeleo, Praha 3: 13–17

CHITTKA, L., A. SHMIDA, N. TROJE & R. MENZEL (1994): Ultraviolet as a component of flower reflections, and the colour perception of Hymenoptera. – Vision Research, 34(11): 1489–1508

CLAUSSEN, J. (1981): Untersuchungen über den Wurmbefall von Landschildkröten und Therapieversuche mit neueren Anthelminthika. – München, Ludwig-Maximilians-Universität, Diss.

CONNOR, M.J. (1993): *Testudo kleinmanni*, notes on captive husbandry. – Tortuga Gazette 29(1): 5–6

CORTON, M. (1996): Understanding common worms and protozoa. TortoiseTrust Website https://www.tortoisetrust.org/articles/parasites.html

D'ALTERIO, G. L. D., A. W. SAINSBURY, S.K. MACGREGOR, S. M. BAHA EL DIN, M. TAYLOR & E.L. WENMAN (1999): Health assessment of the Egyptian tortoise *Testudo kleinmanni* in release studies – Verh. Ber. Erkrg. Zootiere 39: 433–435

DATHE, F. (2003): Pflege und Vermehrung der Ägyptischen Landschildkröte, *Testudo kleinmanni* (LORTET, 1883), im Tierpark Berlin-Friedrichsfelde. – Milu 11(2): 156–169

DELFINO, M., F. CHESI & U. FRITZ (2009): Shell Morphology of the Egyptian tortoise, *Testudo kleinmanni*, LORTET, 1883, the Osteologically Least-Known Testudo Species. – Zoological Studies 48(6): 850–860

DENNERT, C. (2008): Ernährung von Landschildkröten. – Natur und Tier-Verlag, Münster: 136 S.

DEVEAUX, B. (1997): *Testudo kleinmanni*, la discrète. La Tortue (37): 4–13

— (1997): *Testudo kleinmanni*, die scheue Schönheit. Übersetzt von Dr. M. MEYER. – REPTILIA 2(5): 43–51

DIEGEL, I. (2010): Aktuelle Empfehlungen zur UV-Versorgung von Reptilien. – elaphe 3/2010: 26–29

— (2011): Schätzung der UVB-Leistung von Terrarienlampen mit den Solarmetern 6.5 und 6.2 und Aktuelle Lampenempfehlungen – Update 2011. Download der Zusammenfassung des UVB-Vortrages bei der AGARK-Tagung in Bonn am 2.4.2011. Empfehlungen zur UVB-Versorgung von Reptilien. http://tierarzt-schwelm.de/index.php?page=uv-licht-reptilien

DWORSCHAK, C. (2010): Ein Halbwüstenhaus für *Testudo kleinmanni*. – SACALIA 8(26): 6–11

— (2012): Die Ägyptische Landschildkröte, *Testudo kleinmanni*, LORTET, 1883 – die kleinen Schildkröten mit hohen Ansprüchen. – SACALIA 10(34): 12-35

— (2013): Verhalten von *Testudo kleinmanni* bei unterschiedlicher Haltungsform. – SACALIA 11(39): 22–25

EGGENSCHWILER, U. (1997): Zucht der Ägyptischen Landschildkröte (*Testudo kleinmanni*) in Gefangenschaft. – SIGS-Info 6(1): 13–18

— (1998): Ägyptische Landschildkröte (*Testudo kleinmanni*) – 5 nach 12. – SIGS-Info 7(1): 6–8

— (2000): Die Schildkröte in der tierärztlichen Praxis, vom Praktiker für Praktiker. – Schöneck Verlag, Siblingen: 150 S.

— (2008): Hexamiten bei Schildkröten, Newsletter Nr. 8, online PDF http://www.eggenschwiler.swissworld.com/Hexamiten.pdf

Ernst, C. H. & R. W. Barbour (1989): Turtles of the world. – Smithsonian Institution Scholarly Press, Washington D. C.: 313 S.

Ewert, M. A. (1985): Embryology of turtles. – In: Biology of the Reptilia Vol. 14, Development A., Carl Gans (Ed.). – John Wiley & Sons, New York: 778 S.: 75–257

—, D. R. Jackson & C. E. Nelson (1994): Patterns of temperature-dependent sex determination in turtles. – J. experimental Zool. 270: 3–15

— & C. E. Nelson (1991): Sex determination in turtles: divers patterns and some possible adaptive values. – Copeia, Lawrence 1: 50–69

Farkas, B. L. (1993): Life history notes: *Testudo kleinmanni* (Egyptian tortoise). Maximum size. – Herpetological Review 24(3): 104

—, L. Sasvári & J. R. Buskirk (1997): Maximum size of the Egyptian tortoise, *Testudo kleinmanni*. – Chelonian Conservation and Biology, Lunenburg 2(3): 415

Fass, U. (2013): Die Haltung der Steppenschildkröten *Agrionemys horsfieldii*: Faszinierende Tiere aus ariden Gebieten. Kleintierverlag, Biebertal: 183 S

Ferguson, G. W., J. R. Jones, W. H. Gehrmann, S. H. Hammack, L. G. Talent, R. D. Hudson, E. S. Dierenfeld, M. P. Fitzpatrick, F. L. Frye, M. F. Holick, T. C. Chen, Z. Lu, T. S. Gross & J. J. Vogel (1996): Indoor husbandry of the Panther Chamaeleon *Chamaeleo (Furcifer) pardalis*: Effects of dietary vitamins A and D and ultraviolet irradiation on pathology and life-history traits. – Zoo Biology 15: 279–299

—, A. M. Brinker, W. H. Gehrmann, S. E. Bucklin, F. M. Baines & S. J. Mackin (2010): Voluntary exposure of some western-hemisphere snake and lizard species to ultraviolet-B radiation in the field: how much ultraviolet-B should a lizard or snake receive in captivity? – Zoo Biology 29: 317–334

—, Gehrmann W.H., A.M. Brinker & G.C. Kroh (2014): Daily and seasonal patterns of natural ultraviolet light exposure of the western sagebrush lizard (*Sceloporus graciosus gracilis*) and the dunes sagebrush lizard (*Sceloporus arenicolus*). – Herpetologica 70: 56–68

Fischer, O. (2016): Die richtige Beleuchtung für Bartagamen und andere sonnenliebende Reptilien. – X-Reptile Ratgeber 2016/2017, 19 S.

— (2017): Noch mehr Licht; Stellungnahme von Oliver Fischer (Zoologe). – REPTILIA 123, 22(1): 58–60

— (2017/2018): Wie stelle ich die Beleuchtung für mein Terrarium sonnenliebender Reptilien richtig zusammen? – REPTILIA 128, 22(6): 16–27

Flower, S. S. (1921): Exhibition of, and remarks upon, living specimens of *Testudo leithii* and *T. ibera*. In: Proceedings of the General Meetings for Scientific Business of the Zoological Society of London. – Longmans, Green and Co., London: 645

— (1933): Notes on the recent reptiles and amphibians of Egypt. – Journal of Zoology 103(3): 745–749

Fritz, U. & Buskirk, J. R. (1997): The occurence of *Testudo kleinmanni* in Tripolitania. Čas. Nár. Muz., Ř. Přirodověd 166(1-4): 143–144

Fritzsche, D. & M. Fritzsche (2002): Erfahrungen bei der Haltung und Nachzucht von *Testudo kleinmanni*, Lortet, 1883, der Ägyptischen Landschildkröte. – RADIATA 11(3): 31–39

Frye, F. (1991): Biomedical and Surgical Aspects of Captive Reptile Husbandry. Malabas, Krieger Publishing Company. Malabar (Florida): 712 S.

Frynta, D., L. Kratochvil, J. Moravec, P. Benda, R. Dandova, M. Kaftan, K. Klosova, P. Mikulova, P. Nova & L. Schwarzova (2000): Amphibians and reptiles recently recorded in Libya. – Acta Societatis Zoologicae Bohemicae 64: 17–26

GAD, J. (2000): Die Haltung und Zucht der Ägyptischen Landschildkröte *Testudo kleinmanni*. – SALAMANDRA 36(2): 113–120

— (2001): Vorschläge zur Ernährung der Ägyptischen Landschildkröte *Testudo kleinmanni*, LORTET, 1883, in menschlicher Obhut. – EMYS 8(3): 18–34

— (2005): Ergänzende Bemerkungen zur Haltung und Zucht sowie zum taxonomischen Status der Ägyptischen Landschildkröte *Testudo kleinmanni*, LORTET, 1883. – EMYS 12(4): 4–21

GARDINER, D. W., F. M. BAINES, & K. PANDHER (2009): Photodermatitis and Photokeratoconjunctivitis in a Ball Python (*Python regius*) and a Blue-Tongue Skink (*Tiliqua* spp.). – Journal of zoo and wildlife medicine, 40(4): 757–766

GEFFEN, E. (1985): On the biology of the Egyptian tortoise (*Testudo kleinmanni*) in Holot Agur, Irael. – Diplomarbeit, Tel Aviv Universitiy: 83 S.

— (1987): Imprints of activity of the desert tortoise (*Testudo kleinmanni*) in the dunes of Agur. – Hardun J. Israel Herpetological Information Center, 4: 15–26

— (1990): The threatened Egyptian Tortoise (*Testudo kleinmanni*): Proposal for a reserve at Holot Agur (Western Negev, Israel). – Brit. Herpetol. Soc. Bull., 32: 6–8

— & H. MENDELSSOHN (1988). Home range use and seasonal movements of the Egyptian tortoise (*Testudo kleinmanni*) in the northwestern Negev, Israel. – Herpetologica, 44(3): 354–359

— & H. MENDELSSOHN (1989): Activity patterns and thermoregulatory behavior of the Egyptian tortoise *Testudo kleinmanni* in Israel. – Journal of Herpetology, 23(4): 404–409

— & H. MENDELSSOHN (1991): Preliminary study on the breeding pattern of the Egyptian tortoise, *Testudo kleinmanni*, in Israel. – Herpetological Journal, 1(12): 574–577

— & H. MENDELSSOHN (1997): Effects of Military Activities on Tortoises in Israel. In: J. VAN ABBEMA (Hrsg.): Proceedings – Conservation, Restoration and Management of Tortoises and Turtles – An International Conference, New York Turtle and Tortoise Society, New York: 73

— & H. MENDELSSOHN (1997): Avian Predation on Tortoises in Israel. In: J. VAN ABBEMA (Hrsg.): Proceedings – Conservation, Restoration and Management of Tortoises and Turtles – An International Conference, New York Turtle and Tortoise Society, New York: 105

— & H. MENDELSSOHN (1997): Specific Threats to Tortoises in Israel. In: J. VAN ABBEMA (Hrsg.): Proceedings – Conservation, Restoration and Management of Tortoises and Turtles – An International Conference, New York Turtle and Tortoise Society, New York: 321

Gehrmann, W. H. (1994): Light requirements of captive amphibians and reptiles. – In: MURPHY, J. B., K. ADLER & J. T. COLLINS (Hrsg.): Captive Management and Conservation of Amphibians and Reptiles. – Soc. Study. Amphib. Reptiles (SSAR): 53-59

GEISSEL, U. (2015): LED-Beleuchtung in der Terraristik. – TERRARIA/elaphe 6/2015: 14–21

— (2016): Halogen-Metalldampflampen in der Terrarientechnik. – REPTILIA 121, 21(5): 30–37

— & J. KOBER (2015): Terrarientechnik. – Natur und Tier-Verlag, Münster: 258 S.

GIANETTO, S., E. BRIANTI, G. POGLAYEN, C. SORGI, G. CAPELLI, M. G. PENNISI & G. COC (2007): Efficacy of oxfendazole and fenbendazole against tortoise (*Testudo hermanni*) oxyurids. – Parasitol Res 100(5): 1069–1073

GIBSON, R. (1997): Husbandry of *Testudo kleinmanni* at the Jersey Wildlife Preservation Trust. – Informationsblatt des Jersey Wildlife Preservation Trust

— (2001): Husbandry of *Testudo kleinmanni* at the Jersey Wildlife Preservation Trust. – Informationsblatt des Jersey Wildlife Preservation Trust

GRIMM, M. (2016): Die richtige Terrarienbeleuchtung für Chamäleons und andere Reptilien. – REPTILIA 121, 21(5): 18–29

GRUBER, J. (1993): Haltung und Zucht von *Testudo kleinmanni.* – SIGS-Info, Dezember 1993, 4–5

— (1998): Die Ägyptische Landschildkröte (*Testudo kleinmanni*) Lebensweise, Gefangenschaftshaltung und Nachzucht. – Vortrag im Rahmen eines Arbeitstreffens zum Zuchtprogramm bedrohter Schildkröten der Schildkrötengruppe der ÖGH, Wien 1998. Zusammenfassung

GÜNTHER, A. (1869): Report on two Collections of Indian Reptiles. – In: Proceedings of the Scientific Meetings of the Zoological Society of London. – Longmans, Green, Reader and Dyer, London 37(1): 501–507

HÄFELI, W. & P. ZWART (2000): Panzerweiche bei jungen Landschildkröten und deren mögliche Ursachen. – Prakt Tierarzt 81(2): 129–132

HARTMANN, C. M. (2009): Untersuchungen zur Zusammensetzung von Reptilieneiern. – Dissertation, LMU München: Tierärztliche Fakultät: 336 S.

HEIKKILÄ, A., P. KÄRHÄ, A. TANSKANEN, M. KAUNISMAA, T. KOSKELA, J. KAUROLA, T. TURE & S. SYRJÄLÄ (2009): Characterizing a UV chamber with mercury lamps for assessment of comparability to natural UV conditions. – Polymer Testing , 28(1): 57–65

HEMPEL DE IBARRA, N., M. VOROBYEV & R. MENZEL (2014): Mechanisms, functions and ecology of colour vision in the honeybee. – J. Comp. Physiol. A. Neuroethol. Sens. Neural. Behav. Physiol. 200(6): 411–433

HIBMA, J. C. (2004): Dietary Vitamin D3 and UVB Exposure Effects on Green Iguana Growth Rate: Is Full-Spectrum Lighting Necessary? – Bull. Chicago Herp. Soc. 39(8): 145–150

HIGHFIELD, A.C. (1991): Introducing *Testudo kleinmanni.* – Tortoise Trust Newsletter 5(5/6): 15–17

— (1992): Notes on Dietary Constituents for Herbivorous Terrestrial Chelonians and their effect on Growth and Development. – Tortoise Trust Website http://tortoisetrust.org/articles/dietcons.html

— (1994): Effects of Habitat Degradation and Climatic Change Upon the Behavior of Land Tortoises in the Arid Regions of North Africa and the Middle East. In: 18th International Herpetological Symposium, SCHUETT, G. W. (Ed.). – IHS, New Orleans: 3–4

— (1996): 250 Egyptian Tortoises in Cairo Emergency. – Tortoise Trust Newsletter 10(4): 8–9

— (1996): Practical Encyclopedia of Keeping and Breeding Tortoises and Freshwater Turtles. – Carapace Press, London: 295 S.

— (1998): Protozoan Infections in Chelonians. – Tortoise Trust Website https://www.tortoisetrust.org/articles/proto.html

— (2001-2002): Understanding Microclimates in Captivity. – Website Tortoisetrust.org http://tortoisetrust.org/articles/microclimates.htm

— (2003): Worming difficult to handle tortoises and turtles. – Tortoise Trust Website http://www.tortoisetrust.org/articles/worming.html

— (2005): An outline on treating 'RNS' or respiratory diseases in chelonia. – Tortoise Trust Website https://www.tortoisetrust.org/articles/rns.html

— (2001–2008): Understanding Vivarium Substrates. – Tortoise Trust Website http://tortoisetrust.org/articles/substrates.html

— (2010): The causes of "Pyramiding" deformity in tortoises: a summary of a lecture given to the Sociedad Herpetologica Velenciana Congreso Tortugas on October 30 2010. – Tortoise Trust Website http://www.tortoisetrust.org/articles/pyramiding.html

— (2015): Understanding Reptile Lighting Systems. – Tortoise Trust Website http://www.tortoisetrust.org/articles/lighting.html

— & J. MARTIN (1995): Captive Breeding of the Egyptian Tortoise (*Testudo kleinmanni*). – Reptilies 3(3): 48–52

— & J. MARTIN (2001): Captive breeding of the Egyptian tortoise *Testudo kleinmanni.* – Tortoise Trust Website http://tortoisetrust.org/articles/kleinmanni.html

HLAWATSCH, W. (2009): Vorkommen von *Testudo kleinmanni* bei Misurata in Libyen – elaphe N.F. 17(4): 62–65

HOBBS, L.L. & P.A. MAY (1993): A *Testudo kleinmanni* breeding. – Bull. Chicago Herp. Soc. 28(11): 233–23

HOBY, S., C. WENKER, N. ROBERT, T. JERMANN, S. HARTNACK, H. SEGNER, C.-P. AEBISCHER & A. LIESEGANG (2010): Nutritional Metabolic Bone Disease in Juvenile Veiled Chameleons (*Chamaeleo calyptratus*) and Its Prevention.- J. Nutr. 140(11): 1923–1931

HOEKSTRA, R. (1994): Was ich über die Haltung der Ägyptischen Landschildkröte (*Testudo kleinmanni*) lernte. – SIGS-Info, Juni 1994: 8–12

— (1994): What I have learned of the husbandry of the Egyptian tortoise, *Testudo kleinmanni*. – Tortuga Gazette 30: 6–7

— (1996): Was ich über die Haltung der Ägyptischen Landschildkröte (*Testudo kleinmanni*) lernte. – EMYS 3(1): 12–15

HOLLIS, B. W. (2011) Chapter 47 – Detection of Vitamin D and Its Major Metabolites. – In: FELDMAN, D., PIKE, W. J. & J. S. ADAMS (Eds.): Vitamin D (Third Edition). – Elsevier, Amsterdam: 823–844

HUFER, H. (2005): Moderne Lichttechnik im Schildkrötenterrarium. – Schildkröten im Fokus 2(1): 21–28

IBRAHIM, Abdel A. (2001): The reptile community of the Zaranik Protected area, North Sinai, Egypt with special reference to their ecology and conservation. – Egyptian Journal of Natural History 3: 81–92

JACOBS, G. H. (1996): Primate photopigments and primate color vision. – Proc. Nat. Acad. Sci. USA 93(2): 577–581

JACOBSON, E.R., J. M. GASKIN, M. B. BROWN, R. K. HARRIS, C. H. GARDINER, J. L. LAPOINTE, H. P. ADAMS & C. REGGIARDO (1991): Chronic upper respriatory tract disease of free-ranging desert tortoises (*Xerobates agassizii*). – Journal of Wildlife Diseases, 27(2): 296–316

—, M. B. BROWN, L. D. WENDLAND, D. R. BROWN, P. A. KLEIN, M. M. CHRISTOPHER & K. H. BERRY (2014): Mycoplasmosis and upper respiratory tract disease of tortoises: A review and update. – The Veterinary Journal, 201(3): 257–264

JASSER-HÄGER, I. & A. WINTER (2007): Ergebnisse des Inkubationsprojektes für Landschildkröten von 2002 bis 2007. – RADIATA 16(3): 2–41

JENNEMANN, G. (2010): Lithophagie bei Schildkröten – Warum fressen Schildkröten eigentlich immer wieder Steine? Ursachen und Risiken. – Schildkröten im Fokus, Bergheim 7(3): 10–25

KARSTEN, K. B., G. W. FERGUSON, T. C. CHEN & M. F. HOLICK (2009): Panther Chameleons, *Furcifer pardalis*, Behaviorally Regulate Optimal Exposure to UV depending on Dietary Vitamin D3 Status. – Physiological and Biochemical Zoology 82(3): 218–225

KAŹMIERCZAK, M. (2007): Żółw egipski *Testudo kleinmanni* LORTET, 1883. Na czym polega problem jego hodowli? – DRACOmagazyn 2(5): 40–45

KELBER A., M. VOROBYEV & D. OSORIO (2003): Animal colour vision – behavioural tests and physiological concepts. – Biol. Rev Camb. Philos. Soc. 78(1): 81–118.

KIRSCHE, W. (1998): Die Landschildkröten Europas. – Mergus Verlag, Melle: 104 S.

KNAPPE, G. (2011): Vermehrung Europäischer Landschildkröten, Bebrütung der Eier und Anmerkungen zur Anzucht der Schlüpflinge – meine Methode. – Schildkröten im Fokus 8(2): 24–34

KÖHLER, G. (2004): Inkubation von Reptilieneiern: Grundlagen, Anleitungen und Erfahrungen. – Herpeton Verlag, Offenbach: 254 S

KÖHLER, H. (2009): Über Bewegungsradius und Ruhepausen wild lebender maurischer Landschildkröten. – Schildkröten im Fokus 6(1): 29–34

— (2011a): Die UV-Bestrahlung von Landschildkröten: im Biotop und bei der Innenhaltung. – Website http://www.schildi-online.eu/fachartikel 2008–2011

— (2011b): Vorsicht beim Einsatz ultra-starker UV-Strahler. – Schildkröten im Fokus 8(3): 13–16

Kölle, P. (2000): Landschildkröten – Die häufigsten Parasitosen. Kleintier konkret 1: 19–22

—, & M. Pees (2008): Die Schildkröte Heimtier und Patient. P. Kölle (Hrsg.), Enke Verlag, Stuttgart: 270 S.

— & S. Blahak (2015): ReptilienSkills – Praxisleitfaden Schildkröten, Echsen und Schlangen: MemoVet. – Schattauer Verlag, Stuttgart: 334 S.

Kopsch, G. (2006): Untersuchungen zur Körperzusammensetzung von Schildkröten. – Dissertation, LMU München: Tierärztliche Fakultät: 185 S.

Krug, P. (2010): Darmparasitosen Bei Landschildkröten. – PDF-Download: Website http://www.schildkroeten-stammtisch-rhein-main.de/index.php/grundlagen/ersteschildkroete/cat_view/45-?start=10

Kunz, K. (2015): LED-Terrarienbeleuchtung in der Praxis. – TERRARIA/elaphe 6/2015: 36–40

Lapparent de Broin, F., R. Bour & J. Perälä (2006a): Morphological definition of *Eurotestudo* (Testudinidae, Chelonii): first part. – Ann. Paléontol. 92(4): 255–304

—, R. Bour & J. Perälä (2006b): Morphological definition of *Eurotestudo* (Testudinidae, Chelonii): second part. – Ann. Paléontol. 92(4): 325–357

Lindgren, J., W. H. Gehrmann, G. W. Ferguson & J. E. Pinder (2008): Measuring Effective Vitamin D3-Producing Ultraviolet B Radiation Using Solartech's Solarmeter 6.4 Handheld UVB Radiometer. – Bulletin of the Chicago Herpetological Society 43(4): 57–62

Loehr, V. (1999): Behandeling van zeer slechte conditie van wildvang Egyptische landschildpadden (*Testudo kleinmanni*). – De Schildpad 25(3): 120–127

Loew, E. R. & V. J. Govardovskii (2001): Photoreceptors and visual pigments in the red-eared turtle, *Trachemys scripta elegans*. – Vis. Neurosci., 18(5): 753–757

Lortet, L. (1883): Études zoologiques sur la faune du Lac de Tibériade suivies d'un aperçu sur la faune des Lacs d'Antioche et de Homs; Poissons et Reptiles du Lac de Tibériade et de quelques autres parties de la Syrie. In: Archives du Muséum d'Histoire Naturelle de Lyon, Tome Trosième. – H. Georg, Lyon: 188–189

— (1887): Observations sur les tortues terrestres et paludines du bassin de la Méditerannée. – In: Archives du Muséum d'Histoire Naturelle de Lyon, Tome Quatrième. – H. Georg, Lyon: 11–14

Loveridge, A. & E. E. Williams (1957): Revision on the African tortoises and turtles of the suborder of Cryptodira. – Bull. Mus. Comp. Zool. 115(6): 161–557

Macale, D., M. Scalici & A. Venchi (2009): Growth, mortality and longevity of the Egyptian tortoise *Testudo kleinmanni* Lortet, 1883. – Israel Journal of Ecology & Evolution 55 (2): 133–147

—, A. Venchi & M. Scalici (2011): Shell shape and size variation in the Egyptian tortoise *Testudo kleinmanni* (Testudinidae, Testudines). – Folia Zool. 60 (2): 167-175

MacLaughlin, J., R. Anderson & M. F. Holick (1982): Spectral character of sunlight modulates photosynthesis of previtamin D3 and its photoisomers in human skin. – Science 216(4549): 1001–1003

Marschang, R. E., M. Gravendyck & E. F. Kaleta (1997): Herpesviruses in Tortoises: Investigations into Virus Isolation and the Treatment of Viral Stomatitis in *Testudo hermanni* and *T. graeca*. – Journal of veterinary medicine, Series B 44(7): 385–394

—, K. Milde & M. Bellavista (2001): Virus isolation and vaccination of mediterranean tortoises against a chelonid herpesvirus in a chronically infected population in Italy. – Deutsche tierärztliche Wochenschrift 108(9): 376–379

—, & R. M. Schneider (2007): Antibodies to viruses in wild-caught spur-thighed tortoises, *Testudo graeca* in Turkey. – Veterinary Record 161(3): 102–103

—, T. Papp, L. Ferretti, S. Hochscheid & F. Bentivegna (2009): Detection and partial characterization of Herpesviruses from Egyptian tortoises (*Testudo kleinmanni*) imported into Italy from Libya. – Journal of Zoo and Wildlife Medicine 40(1): 211–213

— (2011): Viruses Infecting Reptiles. – Viruses 2011, 3: 2087–2126

Mascort, R. (2009): *Testudo Kleinmanni* in Ägypten: die letzten Überlebenden. – SACALIA 7(24): 22-36

Maślak R. & M. Kaźmierczak (2010): Żółw egipski (*Testudo kleinmanni* Lortet, 1883). Biologia i problemy ochrony. Wydawnictwo Lineart, Warszawa. 96 stron, 17 zdjęć, 124 cytowane publikacje. Verlag Lineart, Warschau: 96 S.

Mathes, K. A. (2003): Untersuchungen zum Vorkommen von Mykoplasmen und Herpesviren bei freilebenden und in Gefangenschaft gehaltenen Mediterranen Landschildkröten (*Testudo hermanni*, *Testudo graeca graeca* und *Testudo graeca ibera*) in Frankreich und Marokko. – VVB Laufersweiler Verlag èdition scientifique, Wettenberg: 279 S.

Matsuoka, Y. (1994): Captive breeding and growth of Egyptian tortoises. – Anim. Zoos. 46(2): 4–7

McLaughlin, G. S. (1997): Upper Respiratory Tract Disease in Gopher tortoises, *Gopherus polyphemus*: Pathology, Immune Responeses, Transmission an Implication for Conservation and Management. – University of Florida, A Dissertation Presented to the Graduate School of the University of Florida in Partial Fulfillment of the Requirements for the Degree of Doctor of Philosophy: 121 S.

Mendelssohn, H. (1982): Egyptian tortoise. – In: The IUCN AmphibiaReptilia Red Data Book, B. Groombridge & L. Wright (Eds.). – IUCN, Gland, Schweiz: 133–136

—, & E. Geffen (1987): The Egyptian tortoise in Israel. – Israel Land and Nature 12(4): 153–157

—, & E. Geffen (1987): Egyptian tortoise needs refuge in Israel. – Oryx 21(4): 250

—, & E. Geffen (1995): The Egyptian tortoise (*Testudo kleinmanni*): 139-145. – In: Ballasina, D. (Hrsg.): Red data Book on Mediterranean Chelonians. – Bologna (Edacricole): 190 S.

Meyer de Rojas, G. (2013): Das Rätselraten um Schildanomalien. – Schildkröten im Fokus 10(2): 25–34

— (2018): http://testudoland.npage.de – Das Rätselraten um Schildanomalien. Und: Inkubation – Was passiert im Ei? Website http://testudoland.npage.de/anomalien.html sowie http://testudoland.npage.de/was-passiert-im-ei.html,

Muftah, A. (2017): Some 1,900 tortoises seized in Benghazi. – Libya Herald, Nachrichtenmagazin online https://www.libyaherald.com/2017/04/27/some-1900-smuggled-tortoises-seized-in-benghazi/#axzz4fYZ1CHdP

Pellitteri-Rosa, D., R. Sacchi, P. Galeotti, M. Marchesi & M. Fasola (2010): Do Hermann's tortoises (*Testudo hermanni*) discriminate colours? An experiment with natural and artificial stimuli. – Italian Journal of Zoology 77(4): 481–491

Perälä, J. (1999): Egyptinkilpikonna, *Testudo kleinmanni*, Lortet, 1883, sukupuuton partaalla – Herpetomania, 1999(4)

— (2001): A new species of *Testudo* (Testudines: Testudinidae) from the Middle East, with implications for Conservation. – J. Herpetology 35(4): 567–582

— (2003): *Testudo kleinmanni*. The IUCN Red List of Threatened Species 2003: e.T21652A9306908 http://dx.doi.org/10.2305/IUCN.UK.2003.RLTS.T21652A9306908.en

— (2005): Assessment of the threatened status of *Testudo kleinmanni* Lortet, 1883 (Testudines: Testudinidae) for the IUCN Red List. – Chel. Cons. Biol. 4(4): 891–899

— (2006): Assessment of the Threatened Status of *Testudo werneri*, Perälä 2001 (Testudines: Testudinidae) for the IUCN Red List – Chelonian Conservation & Biology 5(1): 57–66

PHILIPPEN, H.-D. (1990): Ägyptische Landschildkröte – Aktivitäts- und Thermoregulationsverhalten in Israel. DATZ 12: 714

— (2005): Ägyptische Landschildkröte *Testudo kleinmanni* (LORTET, 1883). – MARGINATA 2(3): 31 – 34

— (2006): Über 400 Schildkröten in Koffer am Mailänder Flughafen. – MARGINATA 3(4): 5

— & C. SERBENT (1986): Zur Biologie und Ökologie von *Testudo* (*Pseudotestudo*) *kleinmanni* – der Ägyptischen Landschildkröte. – Die Schildkröte N.F. 1 (4): 3 – 20

PIEH, A. (2000): Arten und Unterarten der Landschildkröten des Mittelmeergebietes – DRACO 02: 4 – 17

— & H.-D. PHILIPPEN (2007): Mediterrane Landschildkröten. – DRACO 32: 4 – 34

PRAUSER, L. (2013): Calciumernährung (#7): Die Eierschalenfrage. – Onlineartikel http://www.testudowelt.de/?p=4801

PRITCHARD, P. C. H (1979): Encyclopedia of turtles. – London – T. F. H. Publications, Neptune City: 895 S.

— (1985): *Testudo kleinmanni* in Egypt and Israel. – IUCN Tortoise Group Newsletter 1:3

PURGLEY, H., J. JEWELL, J. E. DEACON, R. M. WINOKUR & V. M. TRIPOLI (2009): Vitamin D3 in captive Green Sea Turtles (*Chelonia mydes*). – Chelonian Conservation and Biology 8(2): 161 – 167

RANK, S. J. & E. KAUFMAN (2012): Report of the Analysis of the Wild Diet of *Testudo werneri*. – Onlineartikel PDF http://www.nature-conservation.org.il/BuildaGate5/portals/conserve/imagesP/sub96/960799254.pdf

REINARZ, A. (2008): *Testudo kleinmanni* – Haltung und Nachzucht der Ägyptischen Landschildkröte. – Schildkröten im Fokus 5(4): 3 – 15

REINAUER, S., R. BÖHM & R. E. MARSCHANG (2005): Inactivation of tortoise viruses in the environment. – J. Herp. Med. Surg. 15(3): 10 – 15

ROGNER, M. (1996): Schildkröten 2. – Heidi Rogner Verlag, Hürtgenwald: 265 S.

— (2008): Schildkröten: Biologie, Haltung, Vermehrung – (Datz Terrarienbücher) Verlag Eugen Ulmer, Stuttgart: 160 S.

RUITER, de M. (1994): Ägyptische Landschildkröten. – DATZ 47(12): 753

SABER, A.S. & B. M. KAMAL (2010): Computed Tomography and 3D Reconstruction of the Respiratory Organs of the Egyptian Tortoise (*Testudo kleinmanni*). – Journal of veterinary anatomy 3(1): 1 – 15

SASSENBURG, L. (2005): Schildkrötenkrankheiten. – 3. Aufl., Bede-Verlag, Ruhmannsfelden: 127 S.

SAUER, K., H. SCHUCHART, B. STECK & H.-G. HORN (2004): PraxisRatgeber Vivarienbeleuchtung: Das richtige Licht in Aquarium und Terrarium. – Edition Chimaira, Frankfurt a. M.: 287 S.

SCHEFFEL, R. (2011): Die Ägyptische Landschildkröte, *Testudo kleinmanni* – eine der kleinsten Landschildkröten, aber mit großen Ansprüchen. – Teil 1. – Testudo Welt Magazin 4(11): 37 – 43

— (2011): Die Ägyptische Landschildkröte, *Testudo kleinmanni* – eine der kleinsten Landschildkröten, aber mit großen Ansprüchen. – Teil 2. – Testudo Welt Magazin 4(12): 38 – 43

— (2014): *Testudo kleinmanni*, die Ägyptische Landschildkröte – klein, aber anspruchsvoll. – RADIATA 23(1): 4 – 51

SCHLEICH, H. H. (1984): Reptilien und Amphibien des Kouf National Parkes in NE-Libyen. – LVAR-Mittlg., München 4(1): 9-10 http://lars-ev.de/pdf/Mitteilungen/Mitt_LARS_4-1_1984-03-xx.pdf

— (1987): Contributions to the Herpetology of Kouf National Park (NE-Libya) and Adjacent Areas. – Spixiana, München 10(1): 37 – 80.

— (1989): Merkmalsausbildung an Landschildkröten in Nord-Ost-Libyen (Testudines: Testudinidae). – Herpetozoa, 1 (3/4): 97 – 108

—, W. KÄSTLE & K. KABISCH (1996): Amphibians and reptiles of North Africa. – Verlag: Koeltz Scientific Books, Königstein: 625 S.

SCHMIDT, H.U. (2009): "Naturbruten“ bei der Ägyptischen Landschildkröte, *Testudo kleinmanni*. – RADIATA 18(2): 21 – 25

SCHNEIDER, W. & C. SCHNEIDER (2006): Beobachtungen an der Ägyptischen Landschildkröte, *Testudo kleinmanni*, LORTET, 1883, in Libyen. – RADIATA 15(1): 11 – 19

SCHNEIDER, C. & W. SCHNEIDER (2008): The Egyptian Tortoise, *Testudo kleinmanni* LORTET, 1883 in Lybia. – SALAMANDRA 44(3): 141 – 152

SCHRAMM, R. (2016): Erste Erfahrungen mit LED-Beleuchtung bei Spaltenschildkröten. – Website http://www.testudowelt.de/?p=9870

— (2017): Landschildkröten – Futterpflanzen. – 3. Aufl. Tartaruga-Verlag, Grebenhain: 176 S.

SCHUMACHER, I. M., M. B. BROWN, E. R. JACOBSON, B. R. COLLINS & P. A. KLEIN (1993): Detection of Antibodies to a Pathogenic Mycoplasma in Desert Tortoises (*Gopherus agassizii*) with Upper Respiratory Tract Disease. – Journal of Clinical Microbiology, Vol. 31, 6: 1454 – 1460

SENNEKE, D. M. (2004): Nutrient analysis of replacement turtle and tortoise food. – Website World Chelonian Trust http://www.chelonia.org/Articles/nutrientanalysis.htm

SHACHAM, B. (2004): *Testudo werneri*. – Manouria 7(22): 44 – 45

SHALTOUT, K.H. (2004): Application of the Ecosystem Approach – Mediterranean Case Studies Omayed Biosphere Reserve (OBR), Egypt. – IUCN Center for Mediterranean Cooperation, Contract No. 22/04/Project No. 7333-000 FLI1

SIMANG, A., P. L. CUNNINGHAM & B. T. HENEN (2010): Color Selection by Juvenile Leopard Tortoises (*Stigmochelys pardalis*) in Namibia. – Journal of Herpetology 44(2): 327 – 331

ŠIROKÝ, P & U. FRITZ (2007): Is *Testudo werneri* a distinct species? – Biologia Bratislava 62(2): 1 – 4

SPIER-FOCKENS, E. (1995): Ervaringen met de *Testudo kleinmanni*. – De Schildpad 21: 48 – 51

STEIDELE N. & T. FRIZ (2014): Nachgefragt beim Tierarzt: Vitaminmangel (Hypovitaminose) und Übervitaminisierung (Hypervitaminose). – Website http://www.testudowelt.de/?p=7206

STERNBERG, N. (2017): Reptilien in der tierärztlichen Praxis. Haltungsbedingte Krankheiten sowie Aspekte von Public Health bezogen auf Zoonosen in der Reptilienhaltung. Wissenschaftlicher Aufsatz. – Grin Verlag, München: 32 S.

STETTNER, A. (2004): Haltung und Nachzucht der Ägyptischen Landschildkröte *Testudo kleinmanni*, LORTET, 1883. – EMYS 11(1): 22 – 32

— (2006): Captive maintenance and breeding of the Egyptian tortoise (*Testudo kleinmanni*). In: ARNTER, H., B. FARKAS & V. LOEHR (Hrsg.): Turtles, Proceedings: International Turtle & Tortoise Symposium Vienna 2002. Edition CHIMAIRA Frankfurt am Main: 554 – 563

STIFTUNG WARENTEST (2017): So schneiden gute LED-Lampen im Langzeittest ab. – In: test.de vom 30. März 2017 https://www.test.de/LED-Lampen-So-schneiden-gute-LED-im-Langzeittest-ab-5157021-0/

STUBBS, D. (1989): *Testudo kleinmanni*: Egyptian tortoise. – In: Conservation Biology of Tortoises; Occ. Papers IUCN Species Survival Comm. No. 5: 39 – 40

TABAKA, C. & D. SENNEKE (2003, 2004): Egyptian tortoise – *Testudo kleinmanni*. – World Chelonian Trust Website http://www.chelonia.org/Articles/tkleinmannicare.htm

THOMAS, B. & J. THOMAS (2001): The Egyptian tortoise (*Testudo kleinmanni*). – Tortuga Gazette 37(2): 1 – 2

TRISTRAM, H. B. (1884): The Survey of Western Palestine. The Flora and Fauna of Palestine. – The Committee of the Palestine Exploration Fund, London: 455 S.

UNE, Y., K. UEMURA, Y. NAKANO, J. KAMIIE, T. ISHIBASHI & Y. NOMURA (1999): Herpesvirus Infection in Tortoises (*Malacochersus tornieri* and *Testudo horsfieldii*). – Veterinary Pathology 36(6): 624–627

VELENSKÝ, P. (2002): *Testudo kleinmanni*, zelva egyptská. – AKVÁRIUM – TERÁRIUM – Praha, (5)

VETTER, H. (2002): Ägyptische Landschildkröte – *Testudo kleinmanni* LORTET, 1883. – In: Schildkröten-Lexikon Heft 2, Frankfurt am Main, Edition Chimaira: 32-56

— (2006): Schildkrötenbibliothek 2. Griechische Landschildkröte (*Testudo hermanni*), Dalmatinische und Italienische Landschildkröte – *Testudo boettgeri, hercegovinensis* und *hermanni*. – Edition CHIMAIRA, Frankfurt am Main: 325 S.

VINKE, T. & S. VINKE (2004): Vermehrung von Landschildkröten. – Herpeton Verlag, Offenbach: 189 S

— (2006): Die Vermehrung Europäischer Landschildkröten – Voraussetzungen, mögliche Schwierigkeiten und Lösungen. – In: DAUBNER, M. & T. VINKE (Hrsg.): Testudo – Häufig gehaltene Arten – Schildkröten im Fokus – Sonderband, dauvi-Verlag, Bergheim: 103–116

WEBB, A. R., DECOSTA, B. & HOLICK, M. F. (1989) Sunlight regulates the cutaneous production of vitamin D3 by causing its photodegradation. – Journal of Clinical Endocrinology & Metabolism 68(5): 882–887

WEGEHAUPT, W. (2006): Natürliche Haltung und Zucht der Griechischen Landschildkröte. – Wegehaupt Verlag, Kressbronn: 352 S

— (2009): Naturalistic keeping and breeding of Hermann´s Tortoises – Wegehaupt Verlag, Kressbronn: 352 S

— (2012): Europäische Schildkröten: Lebensraum und Lebensweise. – Wegehaupt Verlag, Kressbronn: 448 S

— (2019 a): Feeder Plants for Mediterranean Tortoises – Wegehaupt Verlag, Kressbronn: 224 p.

— (2019 b): Mediterranean Tortoises, Where and How they Live in the Wild – Wegehaupt Verlag, Kressbronn: 224 p.

WENMAN, E. (2001): Population and habitat status of the kleinmann's tortoise, *Testudo kleinmanni* – species conservation efforts in Egypt. – Proceedings of the International Congress on Testudo Genus

—, L.G. D'ALTERIO & S. M. BAHA EL DIN (2001): Health care assessment of the *Testudo kleinmanni* – an aid to rehabilitation and future reintroduction. Paper presented at Proceedings of the International Congress on Testudo Genus

WERNER, F. (1909): Reptilien, Batrachier und Fische von Tripolis und Barka. – Zool. Jahrb., Syst. Abt. 27(6): 569-646

WERNER, Y. L. (1973): The Reptiles of the Sinai Peninsula. Hebrew Universitiy of Jerusalem

— (1982): Herpetofaunal survey of the Sinai Peninsula (1967–77), with emphasis on the Saharan sand community. – U.S.F.W.S., Wildlife Research Report 13: 153–161

— (1987): Ecological zoogeograpy of the Saharo-Arabian, Saharan and Arabian reptiles in the sand deserts of southern Israel. – In: F. KRUPP, W. SCHNEIDER & R. KINZELBACH (eds.), Proceedings of the Symposium on the Fauna and Zoogeography of the Middle East, Mainz 1985: 227–237, Wiesbaden.

— (1988): Herpetofaunal survey of Israel (1950-85), with comments on Sinai and Jordan and on zoogeographical heterogeneity. p. 355–387. In: Y. YOM-TOV & E. TCHERNOV (eds.), The Zoogeography of Israel: Dordrecht: 600 S.

WIESNER, C. S. & C. IBEN (2003): Influence of environmental humidity and dietary protein on pyramidal growth of carapaces in African spurred tortoises (Geochelone sulcata). – Blackwell Verlag, Berlin. J. Anim. Physiol. a. Anim. Nutr. 87: 66–74

WILLIAMS, E. (1953): Fossils and the Distribution of Chelyid Turtles. – BREVIORA, Mus. of Comp. Zool., Cambridge Mass. 13: 1–11

WILSON, D. S., C. R. TRACY & C. R. TRACY (2003): Estimating age of turtles from growth rings: a critical evaluation of the technique. – Herpetologica 59(2): 178–194

WITTMUND, P. (1999): Haltungsbericht über *Pseudotestudo kleinmanni* – Ägyptische Landschildkröte. – Schildkröten 6(4): 26-27

— (2005): Haltung und Nachzucht von *Testudo kleinmanni*. – elaphe N.F. 13(2): 33–34

WUNDERLICH, S. (2012a): Der Einsatz von UV-Messgeräten in der Terraristik. – TERRARIA/elaphe 5/2012: 26–39

— (2012b): Farbtemperatur und Farbwiedergabeindex – Sinnvolle Größen in der Terraristik? – Terraria/elaphe 5(2012): 40–47

— (2015): LED's – wirklich die bessere Alternative in der Terrarienbeleuchtung? – TERRARIA/elaphe 6/2015: 26–34

— (2016): (UV)-Licht in der Krokodilhaltung. – Vortrag AG Krokodile München Juni 2016

— (2017): Noch mehr Licht – Stellungnahme. – REPTILIA 123, 22(1): 54–58

— (2018): www.licht-im-terrarium.de Rev. 2014/03/08 – 2018/02/04

ZAVATTARI, E. (1930): Erpetologia della Cirenaica. – Archivio Zoologico Italiano 14: 263

ZINNER, H. (1974): On distribution and population dynamics of snakes in the Negev desert and Sinai. – Israel J. Zool. 23(3/4): 216

ZWAARTEPORTE, H. (2004a): Studbook Breeding Programme *Testudo kleinmanni* – Egyptian or Kleinmann's tortoise. – Annual Report 2003. – European Studbook Foundation PDF https://www.studbooks.eu/site/assets/files/1301/testudokleinmanniannualreport2003.pdf

— (2004b): *Testudo kleinmanni* – Manouria 7(22): 25–26

— (2006): Studbook Breeding Programme Testudo kleinmanni – Egyptian or Kleinmann's tortoise. Annual Report 2004 and 2005. – European Studbook Foundation PDF https://studbooks.eu/site/assets/files/1351/testudokleinmanniannualreport20042005.pdf

— (2007a): Studbook Breeding Programme *Testudo kleinmanni* – Egyptian or Kleinmann's tortoise. Annual Report 2006. – European Studbook Foundation PDF https://www.studbooks.eu/site/assets/files/1368/annualreporttestudokleinmanni2006.pdf

— (2007b): Die Ägyptische Landschildkröte (*Testudo kleinmanni*): Sehr beliebt und kritisch bedroht. – DRACO 32: 74–78.

— (2011): Studbook Breeding Programme Testudo kleinmanni – Egyptian or Kleinmann's tortoise. Annual Report 2010. – European Studbook Foundation PDF https://studbooks.eu/site/assets/files/1547/microsoft_word_-_testudo_kleinmanni_blind_copy_sparks_2010b.pdf

— (2013): Studbook Breeding Programme *Testudo kleinmanni* – Egyptian or Kleinmann's tortoise. Annual Report 2010-2012. – European Studbook Foundation PDF https://www.studbooks.eu/site/assets/files/1666/testudo_kleinmanni_annual_report_2010-2012.pdf

— (2015a): Studbook Breeding Programme *Testudo kleinmanni* – Egyptian or Kleinmann's tortoise. Annual Report 2015. – European Studbook Foundation PDF https://www.studbooks.eu/site/assets/files/1649/testudo_kleinmanni_esf_annual_report_2015.pdf

— (2015b): Captive breeding the Critically Endangered Egyptian tortoise *Testudo kleinmanni* LORTET, 1883, for an in situ recovery project in Egypt. – International Zoo Yearbook 49(1): 42–48

— & R. VERHOEKS (2014): Studbook Breeding Programme *Testudo kleinmanni* – Egyptian or Kleinmann's tortoise. Annual Report 2012-2013. – European Studbook Foundation PDF https://studbooks.eu/site/assets/files/1446/annual_report_2012-2013_testudo_kleinmanni.pdf

Internet Sources

www.chelonia.org

http://christinedworschak.wixsite.com/kleinmanni

www.flowersinisrael.com

www.jaeger-bruttechnik.de

www.licht-im-terrarium.de

www.mediterrane-landschildkroeten.de

http://www.nature-conservation.org.il

http://www.schildi-online.eu

https://studbooks.eu

http://testudoland.npage.de

http://www.testudowelt.de

http://tierarzt-schwelm.de

www.tortoisetrust.org

http://www.uvguide.co.uk

https://www.wikipedia.de/

www.wildflowers.co.il

Image Sources

S. 12: By Unknown– Proceedings of the Zoological Society of London (vol. 1869, page 503), Gemeinfrei, https://commons.wikimedia.org/w/index.php?curid=15499739

S. 15: By Daderot – Own work, CC0, https://commons.wikimedia.org/w/index.php?curid=61800241, Gemeinfrei

S. 16: Left: By Faendalimas – Own work, CC BY-SA 3.0, https://commons.wikimedia.org/w/index.php?curid=9739705

S. 16: Right: By Faendalimas – Own work, CC BY-SA 3.0, https://commons.wikimedia.org/w/index.php?curid=9739768

S. 41: This image is in the public domain because it is a screenshot from NASA's globe software World Wind using a public domain layer, such as Blue Marble, MODIS, Landsat, SRTM, USGS or GLOBE.
By Eric Gaba (Sting – fr:Sting) – Own work; Raster background map: screenshot from NASA World Wind (Public Domain), Public Domain, https://commons.wikimedia.org/w/index.php?curid=2271929

S. 43: By Hatem Moushir – Own work, CC-BY-SA 4.0, https://commons.wikimedia.org/w/index.php?curid=62621919

S. 44: By Stefano Alcini – Own work, CC BY-SA 3.0, https://commons.wikimedia.org/w/index.php?curid=1663846

S. 48: By Hatem Moushir – Own work, CC-BY-SA 4.0, https://commons.wikimedia.org/w/index.php?curid=62621918

S. 49: By Hatem Moushir – Own work, CC-BY-SA 4.0, https://commons.wikimedia.org/w/index.php?curid=62621514

S. 55: Top: By Matt Lavin from Bozeman, Montana, USA – Wyoming big sagebrush steppe: NW of Mammoth UT, CC BY-SA 2.0, https://commons.wikimedia.org/w/index.php?curid=44493274

S. 55: Bottom: By Vassilii Khachaturov – Own work, CC BY-SA 3.0, https://commons.wikimedia.org/w/index.php?curid=31036351

S. 56: From top to bottom: De Gideon Pisanty (Gidip) – Trabajo propio, CC BY 3.0, https://commons.wikimedia.org/w/index.php?curid=3831548

By Kristian Peters – Fabelfroh 15:01, 1 December 2005 (UTC) – own work, CC BY-SA 3.0, https://commons.wikimedia.org/w/index.php?curid=442320

By Denis Barthel (talk) – Own work, CC BY-SA 3.0, https://commons.wikimedia.org/w/index.php?curid=10639266

By MurielBendel – Own work, CC-BY-SA 4.0, https://commons.wikimedia.org/w/index.php?curid=44824547

By The original uploader was Jeantosti at French Wikipedia – Transferred from fr.wikipedia to Commons by Bloody-libu using CommonsHelper., CC BY-SA 3.0, https://commons.wikimedia.org/w/index.php?curid=20387350

By Denis Barthel from German Wikipedia, CC BY-SA 3.0, https://commons.wikimedia.org/w/index.php?curid=26265530

S. 57: From top to bottom: By MrPanyGoff – Own work, CC BY-SA 3.0, https://commons.wikimedia.org/w/index.php?curid=26530503

By Winfried Bruenken (Amrum) – Own work, CC BY-SA 2.5, https://commons.wikimedia.org/w/index.php?curid=1022097

By Gideon Pisanty (Gidip) יטנזיפ ןועדג – Own work, CC BY 3.0, https://commons.wikimedia.org/w/index.php?curid=6703423

By Krzysztof Ziarnek, Kenraiz – Own workk, CC-BY-SA 4.0, https://commons.wikimedia.org/w/index.php?curid=58098701

By Ariel Palmon – Own work, CC BY-SA 3.0, https://commons.wikimedia.org/w/index.php?curid=14520530

By Krzysztof Ziarnek, Kenraiz – Own work, CC BY-SA 4.0, https://commons.wikimedia.org/w/index.php?curid=58105451

S. 80: By Der ursprünglich hochladende Benutzer war Degreen in der Wikipedia auf Deutsch-Improved Baba66 (opt Perhelion) on request;En. translation LocustaFr. translation Eric BajartNl. translation BoH – Übertragen aus de.wikipedia nach Commons; CC BY-SA 2.0 de, https://commons.wikimedia.org/w/index.php?curid=10287551

S. 82: Top: Von Cone-response.svg: w:User:DrBob and w:User:Zeimusuderivative work: Sgbeer – Cone-response.svg (Vectorized version of the GFDL image Cone-response.png uploaded by User:Maxim Razin based on work by w:User:DrBob and w:User:Zeimusu.), CC BY-SA 3.0, https://commons.wikimedia.org/w/index.php?curid=17729332

S. 82: Bottom: By L. Shyamal – Own work, Gemeinfrei, https://commons.wikimedia.org/w/index.php?curid=6308626

S. 113: By © Hans Hillewaert /, CC BY-SA 3.0, https://commons.wikimedia.org/w/index.php?curid=4006233

The authors of all other illustrations are named there directly.

Month / Parameters	July	August	Septen
Hours of day-light	14 h	14 h	13–12 decreasin the mo
Hours of sunshine, heat lamp	12 h	11.5 h	9 h
Temperature day / night	33–35 °C increasing over the month 24 °C	35–38 °C increasing over the month 25–26 °C	35–32 decreasin the mo 24 °C
rel. humidity day / night	50–60 % 65–75 %	50–60 % 65–75 %	50–60 65–75
Rain showers	0	0	first sho the last
Feed	no fresh feed, dry feed freely available	no fresh feed, dry feed freely available	no fresh dry feed availa

ber	October	November	December
h g over nth	12–11.5 h decreasing over the month	11–10.5 h decreasing over the month	10 h
	8 h	7 h	6 h
°C g over nth :	30–26 °C 25–22 °C decreasing over the month	24–26 °C 17–20 °C	18–20 °C 10–15 °C
% %	55–65 % 70–80 %	55–65 % 70–80 %	55–65 % 70–80 %
ver in veek	2–3 per month	1–2 per week	2–3 per week
feed, reely le	Per week: fresh 1st: 1 x 2nd: 2 x 3rd: 3 x 4th: a little daily	fresh feed daily, dry feed freely available	fresh feed daily, dry feed freely available